8/19

WADING RIGHT IN

Wading Right In

Discovering the Nature of Wetlands

Catherine Owen Koning

and Sharon M. Ashworth

with illustrations by
Catherine Owen Koning

The University of Chicago Press
Chicago and London

The University of Chicago Press, Chicago 60637
The University of Chicago Press, Ltd., London
© 2019 by Catherine Owen Koning and Sharon M. Ashworth
Published 2019
Printed in the United States of America

28 27 26 25 24 23 22 21 20 19   1 2 3 4 5

ISBN-13: 978-0-226-55421-1 (cloth)
ISBN-13: 978-0-226-55435-8 (paper)
ISBN-13: 978-0-226-55449-5 (e-book)
DOI: https://doi.org/10.7208/chicago/9780226554495.001.0001

Library of Congress Cataloging-in-Publication Data

Names: Koning, Catherine Owen, author. | Ashworth, Sharon M., author.
Title: Wading right in : discovering the nature of wetlands / Catherine Owen Koning
and Sharon M. Ashworth ; with illustrations by Catherine Owen Koning.
Description: Chicago ; London : The University of Chicago Press, 2019. |
Includes bibliographical references and index.
Identifiers: LCCN 2018060151 | ISBN 9780226554211 (cloth : alk. paper) |
ISBN 9780226554358 (pbk. : alk. paper) | ISBN 9780226554495 (e-book)
Subjects: LCSH: Wetlands—United States.
Classification: LCC GB624 .K655 2019 | DDC 577.68—dc23
LC record available at https://lccn.loc.gov/2018060151

♾ This paper meets the requirements of ANSI/NISO Z39.48-1992 (Permanence of Paper).

Contents

Preface

The wetland literature is awash in textbooks, reference books, guidebooks, and philosophical treatises. Advocates, scientists, and consultants have numerous web-based tools and information at their fingertips. But the general public has few resources to turn to for a good wetland read—a rich story that makes the reader laugh, wonder, and learn.

Recent studies demonstrate that people are more deeply moved by stories than by statistics (Small, Loewenstein, and Slovic 2007). With this in mind, we have gathered the real-life tales of a number of wetland scientists, explorers, and advocates and incorporated those stories into this book. Our goal is for you to learn about wetlands not from a checklist of characteristics but by immersion in a description of real events happening to real people. Through the art of storytelling, we hope to put into your hands the science of wetland ecology and the passion of those who wade into the muck. Each story becomes a portal through which you will visit the wetland and discover its secrets, while also learning important ecological lessons.

Our book is organized by generalized wetland types, all of which are geographically wide-ranging. Wetlands are complex ecosystems, classified by an impressive number of different methods, but the typology we have chosen—based on dominant vegetation—is the one that is most visible and therefore most comprehensible. Each category of wetland is found in almost every state in the United States and many of the Canadian provinces, apart from salt marshes and tidal freshwater marshes, which of course occur in coastal areas. Wetlands particular to certain regions — such as the Louisiana bayous or the Piedmont pocosins — we do not attempt to describe, but the ecological principles outlined in this book do very much apply to these and every other kind of wetland. However, because our experiences in wetlands took place primarily in the Midwest and Northeast of the United States, most of our stories and descriptions come from these regions.

We portray each wetland type through the tales of people who work in these wetlands; along the way, you will understand the driving forces that create wetland conditions, discover the many cool adaptations and structures that form in response to these conditions, and grasp the ecosystem services, or "functions," of each wetland type. We have made a serious attempt to verify that the features and functions we describe are supported by the preponderance of evidence found in the scientific literature and do not reflect just one study or one location, unless specifically noted.

Despite the fact that half of the wetlands in the lower forty-eight states have been drained, filled, or irrevocably altered, we have tried to write an optimistic book. Wetlands are still being destroyed at an alarming rate, but there are many people working to reverse this trend. Chapter 8 describes the inspirational work of the field of wetland restoration — bringing wetlands back! These exciting endeavors restore not only the ecosystem but our faith in humans' capacity for solving problems. The restoration of salt marshes and wet meadows also presents excellent opportunities for confronting the challenges of climate change.

Our last chapter takes a sober look at wetland loss, restoration, and protection, and draws together some of the themes that run through the book. First, as you will soon learn, wetlands are intricately bound with the health of the land and thus our own well-being. Long regarded as nothing more than breeding factories for disease-carrying mosquitoes, wetlands in fact protect our water quality by killing pathogens, degrading pesticides, and converting harmful fertilizer runoff into ordinary components of air. Wetlands also play a critical role in the long-term uptake and storage of the greenhouse gases that create climate change. This theme unfolds within each successive chapter, as every wetland type plays a different role in supporting our material conditions.

Second, the creatures — whether finned or furry, slimy or green — that inhabit these magical places are nothing short of miraculous. Their unexpected adaptive responses to the often harsh conditions of salt marshes, swamps, and other wetlands would challenge the imagination of the wildest science fiction writer.

The final theme of the book is about the people: the bog walkers, swamp stompers, river rats, and marsh haunters who delight in detangling the intricacies of connections among wetland soil, water, microbes, flora, and fauna. They are truly a breed apart, tougher than most, and by necessity endowed with a rich sense of humor — sometimes the only way to get yourself unstuck from the muck is to be able to laugh at the situation. Perhaps the swamp gas seeps into their souls, for these explorers are deeply committed to the wetlands in general, and their special corners of the landscape in particular. Through their eyes, we come to know why the storytellers in this book have devoted their lives to understanding and protecting these special ecosystems. Through their stories, we come to a deeper appreciation of how we must connect to the earth, of the ethical obligations we carry, and how we can reciprocate for all it gives us.

This book is not a textbook and will not cover every facet, function, or feature of every hydrogeomorphic category of wetland. We focus on the fun and the fundamental. Yes, scientists and laypersons alike are subject to exclaiming "Cool!" when discovering such things as moss animals, beaver-fighting trees, underwater spiders, heat-producing plants, and rare shrimp in a clover field. We want you to look at wetlands in a whole new way, to make an emotional connection with the creatures and currents within and, optimally, to cherish and protect these unique places.

INTRODUCTION

Sun Turtles and Superstorms

As Hurricane Carol, a Category 3 storm, bore down on southern New England, people in the region rushed to board up windows, fill bathtubs with water, and stock up on food. It was August 31, 1954, and Carol came ashore on Long Island, New York, bearing winds gusting to 125 miles per hour as it slammed into Long Island, New York—the most destructive storm to hit the area since the hurricane of 1938. After sweeping across Long Island, Carol made landfall again at Old Saybrook, Connecticut, just after high tide and left a path of devastation. Metal-gray darkness swirled, and slanted silver torrents of rain pelted the earth, while screaming hundred-mile-per-hour winds stripped leaves off the trees and tossed bikes, sheds and boats in all directions. Most people hunkered down in candlelit rooms and worried about branches falling on their house and water flooding their basement. The thoughts of at least one child, however, were elsewhere. In the central Connecticut village of Moodus, eight-year-old Frank Golet was worrying about his sun turtles: How would they fare in all this whirling wind and water?

Sun turtles—the local name for the spotted turtle—are often found warming themselves in patches of sunshine on half-rotten logs protruding from the water, or on clumps of sphagnum hummocks on the edges of grassy marshes, mossy bogs, or tangled swamp thickets. These small turtles, usually no bigger than six inches, shine like living jewels, their glossy black carapace flecked with small, elongated oval spots of orangey-yellow. Cuter than any Disneyesque techno-creature, a spotted turtle sighting elicits cries of wide-eyed delight from observers of any age.

Finding sun turtles in wetlands on his grandfather's farm was one of young Frank's favorite activities. Crossing a field, passing Jack's Pond, he'd arrive at a stone culvert that formed a bridge over a watery ditch to search for turtles. The "ditch" was a drainage channel, dug between the field and a red maple swamp, running north to a headwaters stream that fed the Moodus River. After systematically studying the water's surface and the edges of the ditch from both sides of the culvert, he'd carefully survey the bottom, trying to differentiate between quartz pebbles, dappled sunlight, and turtle spots. To avoid alerting his quarry and allowing them to elude capture, Frank would first try to locate the turtles without going into the water. However, if the water was deep, the turtles could be very hard to see. So, in he would wade, slowly and quietly walking upstream through the water, scanning for spotted turtles. Often he'd find them along the sides of the channel, on turtle-size shelves; like mink and muskrats, these turtles often hole up just above the waterline, to rest and watch the world go by.

As soon as Hurricane Carol's wind and rain let up, Frank donned his seventeen-year-old brother's hip waders and raced out to check on his sun turtles. Clomping through the flooded hayfields in his too-big waders, he was thrilled to see numerous turtles swimming in the murky water around the ditch. "There were turtles everywhere—probably flooded out from the bank dens, the pond, and the nearby wetlands, which were all connected. I had the best collecting day of my life!" Frank caught twenty-seven turtles that summer, keeping them in a chicken-wire pen equipped with a washtub full of water, rocks for sunning, and a wooden ramp up to the tub rim. He fed them a balanced diet of raw hamburger and lettuce. "When you are a kid and you have something precious to you, you think if you have it around you all of the time, it is somehow more special," he explained. "In late fall, before ice formed on the water, I returned the turtles to the wetlands so that they could spend the winter in the mud." The turtles—and, more important, the wetlands that supported them—played an important role in Frank's life.

I distinctly remember, around the age of five or six, walking into the maple swamp and looking up at the cinnamon fern canopy (of course, I didn't know its name at the time). Entering the swamp in summer was like entering a dreamland—shady; squishy underfoot, with small, scattered pools of water; skunk cabbage; shrubs growing on mounds, where the trees also stood; birds singing; and the aroma of sweet pepperbush flowers in the air. Every step, every turn, was a new and exciting adventure. The marsh located between the swamp and Jack's Pond was my favorite place to look for the nests of marsh birds, including red-winged blackbirds, swamp sparrows, and common yellowthroats. One day, as I stood between sedge tussocks in the marsh, I felt the ground move beneath my feet and immediately jumped to one side. I then realized that I had been standing on the back of a large snapping turtle mostly buried in the mud! Throughout my life, wetlands have been not only beautiful, fascinating, inspiring places, but sites of great comfort and serenity as well. Sitting on a mound in a mature Atlantic white cedar forest, with sunlight filtering through the branches, can be like sitting in an empty cathedral at dawn. I think I imprinted on wetlands at an early age. (Frank Golet, professor emeritus, University of Rhode Island)

People interact with wetlands in many different ways. For some, the dense vegetation and wet ground are too much to push through from the upland, a perceived barrier to passage. Those lucky enough to be given passage by way of a boardwalk or dry path may see only a tangle of plants or be treated to a flush of birds. Others encounter wetlands from the water—the thicket of dusty green bluejoint grass and neon green bur-reed seen from the seat of a canoe, or the floating mat of sphagnum moss and pink orchids glimpsed between casts of a fishing rod. Each gets to the edge and peers in, wondering what goes on in there.

One could turn to a wetland textbook or a state division of natural resources publication for a description of a red maple swamp. There you would find the characteristics that define the wetland type, lists of attendant flora and fauna, and a statement as to the services the wetland provides. While such information remains important and helpful, it is in Frank Golet's words that we find more than a report; we find ourselves placed in the swamp—

seeing, feeling, and experiencing the swamp. Frank's personal portrayal of this particular wetland lists flora and fauna, but goes further to inspire exploration and elicit caring. Such is the intent of this book—to teach, to energize, and perhaps even to motivate. We wish to convey the fascination and passion for wetlands that those who study and protect them feel when they pause between soil samples, plant surveys, and monitoring wells.

Of course, those pauses may arise from equipment failure, muck-stuck boots, or vanished trails—all common occurrences for wetland explorers. A long, hot day pondering the differences among innumerable strikingly-similar-as-the-day-goes-on plants might not provoke wonder in the moment or inspire poetry, but the ensuing stories are well worth the camaraderie formed in their telling.

Such was the tale from Rob Atkinson of Christopher Newport University in Virginia. Rob, a plant biologist, gave us a twist on the classic "stuck" story. It was just one of those days, a hot, humid summer day, and it remained Rob's job to follow the tape-measure line laid across the Virginia tidal marsh and catalog the plants along a salinity gradient from fresh water to brackish water. This was no stroll through the field, but a slow, mud-sucking slog. Sweet Hall Marsh, where Rob found himself on this particularly long, hot day, is at the fresher end of the tidal wetland spectrum and nestled in the bend of the Pamunkey River; it is a favorite neighborhood of muskrats. As muskrats execute their daily rounds, they make unstable trails through the marsh muck, trails often obscured by vegetation. These submerged trails have an almost gravitational pull to them, making it ridiculously easy to slide into—which is exactly what happened as Rob oozed his way down the plant sampling line toward the end of the day. His left leg sunk into the hidden channel and his right remained stranded in the surface sediments. Decidedly stuck, and despite his somewhat twisted position, Rob resumed his work of cataloging plants before attempting to pry himself from the muck, because that is what field biologists do. He was in this uncomfortable position for a while, duly marking down his plant counts, when he noticed the eyes: two copper eyes staring at him from the channel. Being a plant biologist rather than a wildlife biologist, Rob was unsure of what manner of mud creature now stalked him from less than two feet away—poisonous snake? Giant muck frog? His brain, dehydrated and sun-addled, seized and focused on those beady eyes. Sweat dripped, panic swelled, and before too long, he realized that there were *two* pairs of eyes, another just below the first. Two of these mud-dwelling creatures? Were they mating? Staging a group attack? Well, he thought, this is just a bit too odd. His tunnel vision lessened, and when a greater portion of his world could be made sense of, Rob realized that he was, in fact, being stared down by—his own left boot.

While few of us have been stalked by our own footgear, everyone knows the adrenaline rush of fear as well as the endorphin flood of joy that originates from the exciting moments in any life—the moments that create stories, later told with gusto and made more extreme in each retelling. In this book, we have collected stories of all kinds to convey some of the important and interesting aspects of the ecology and environment of wetlands: Frank's recollection of the birds and plants in a red maple swamp, and Rob's depiction of the soil and character of a tidal marsh. Surrounding these stories is a wealth of data, studies, and experiences focused on not only understanding but saving these varied, enigmatic, and at times equal parts blissful and frustrating systems. Broad groups of people now appreciate how wetlands support the creatures, landscapes, and human needs we take for granted. It was the aftermath of another hurricane that helped many understand the cost of losing wetlands.

The year is 2012. It is October 29, the tail end of the hurricane season. Superstorm Sandy is eyeballing some of the most populated areas in the United States, from New Jersey to Connecticut. Over a thousand miles wide, the storm was responsible for creating high winds in areas as far apart as Bermuda and Wisconsin. It sent nine feet of water ashore in Sandy Hook, New Jersey, and between four and five feet in most areas around New York and Connecticut. The front end of the hurricane hit at high tide, making the storm surge much more dangerous. Adding the storm surge to the normal high tide yielded a twelve-to-fifteen-foot wall of water coming ashore between 8 and 10 p.m. As much as twelve inches of rain fell in some areas, and winds sustained speeds of thirty to fifty miles per hour, gusting to ninety miles per hour.

In the midst of evacuation, of hauling plywood boards that almost carried people off in the high winds, of fastening down anything that could become airborne, of ensuring that all staff had enough time to get out of the evacuation zone, managers and scientists at the Wetlands Institute in Stone Harbor, New Jersey, wondered how their own turtles would fare in the storm.

Specifically, diamondback terrapins—their dark, textured shells are ornamented with bright yellow-orange diamond shapes on the top and a varied pattern of yellow and black on the bottom, contrasting beautifully with the dark-gray speckles covering their legs and body. The marshes around the Wetlands Institute harbor dozens of these rare turtles, creatures subject to death by overhunting, wayward crab traps, and car tires as well as habitat loss.

These small salt marsh turtles are not only stunning to look at, but are also a marvel of evolution. "The terrapins are designed to live in the salt marshes—to deal with changes in salinity, changes in temperature, changes in water level," remarks Lisa Ferguson, director of Conservation and Research at the Wetlands Institute. Terrapin protection is an important part of the institute's work. Lisa's research team conducts surveys, puts up roadside barriers to keep the females from crossing, works with commercial fishing groups to convince them to use crab traps that have an escape hatch for turtles, and even lobbies to reduce turtle hunting. When females seeking to lay their eggs in a cozy upland spot are hit and killed by a car, Lisa's crew will remove the unbroken eggs, incubate them, and rear the hatchlings in captivity for a year before releasing them in what is known as "head-starting." "When we release turtles in our head-starting program"—often employing lucky young visitors to carry the baby turtles back to the edge of the water—"we mark them, and then seven to eight years later, when they are mature, they come back home, to the place they were released, and they generally come back every summer. We look for them, and if they are marked that means they were hatched from our incubators, went out to the marsh and made a living, and then returned when they were mature to lay their own eggs. Some of them have been here for thirteen years or more."

So the institute has a lot invested—time, energy, sweat, heart, as well as dollars—in these terrapins. But how would the diamondbacks deal with a huge storm like Hurricane Sandy? Would the salt marsh protect them from the rushing waters and swirling tides of a hurricane?

Fifty-eight years after Hurricane Carol, the US coastline from Virginia to Maine has changed significantly. Hundreds of thousands of wetland acres have been lost, mostly to new development. Homes, roads, schools, whole communities exist where none stood before, replacing sand bars, reefs, and wetlands. The protective ability of natural ecosystems like wetlands is now so well acknowledged that recovery and rebuilding efforts call for "green infrastructure" or "soft defenses" instead of the hard infrastructure of sea-walls, floodgates, and other large-scale engineering solutions. After Hurricane Sandy, millions of federal dollars have been spent on these kinds of efforts — using nature's own design to protect humans and other creatures. In Jamaica Bay, New York, for example, volunteers worked with the Army Corps of Engineers to create sand islands and replant them with seagrasses. Oyster reefs, once plentiful off the island of Manhattan, are being recreated to absorb incoming wave energy. Up and down the coasts of New Jersey, Connecticut, and Rhode Island, salt marshes are being restored in places they had been destroyed, and the sediment washed away by Sandy is being added back.

But was this work too little, too late to protect the terrapins on Cape May from the force of Sandy? According to Lisa Ferguson, "The salt marshes around the Wetlands Institute are pretty extensive and resilient, and fared pretty well, although other marshes in the area were more extensively impacted. And many of the nesting areas were affected. However, we did not see any drastic changes in the population after the hurricane, although some other terrapin populations in other areas were affected." Just like every year, some of the same terrapins — marked and released in the head-starting program — were found again in the salt marshes that the institute has helped to protect. "To see them returning again, year after year, it's like seeing old friends," Lisa remarked happily.

Extreme events have a way of sharpening our focus, turning our gaze to the problems that sit right in front of us. The wetlands we fill to create developable lands, or flood to make ponds, or poison to use as dump sites, are helping us survive. By transforming toxins and taking the brunt of angry seas, wetlands have been our protectors.

To some people, wetlands present only a blockade — both physically and mentally. They are a place to circumnavigate, an obstacle to overcome, a jungle to struggle through from the firm footing of upland terrain to the stream, river, or lake beyond. They get in the way, they block the path, they are neither open water nor dry land. You can't build on them, and you can't swim in them. They are the places in between, the borders of the lake, the low spots collecting water between hills. They are the places often disregarded completely or altered unrecognizably, made into more familiar and "useful" kinds of terrain. To many, wetlands are just *in the way*.

But to those who pass easily through their borders, who wander through the glorious browns and greens of shrubs and grasses, who bounce on the squishy mounds and breathe in the rich earthy atmosphere, wetlands are holy places. Places where a child can turn over a decaying log and find a dusky salamander; where yellow-necked Blanding's turtles hide in the underwater tangles of meadowsweet; where sunlight beams through red maples, laying stripes of color across the blue-gray back of a heron; where a person can wander a damp and verdant path to the green-blue water of a river's edge. To these people and these creatures, wetlands are not *in* the way, they *are* the way: the way to peace, to beauty, to a strong attachment to a complex evolutionary network on a planet destined not for destruction but for celebration. Storytellers, swamp walkers, turtle watchers—all will wade right in, revealing the richness of the wetlands, letting us in on the secret, so we too can grasp a richly colored, secure future.

CHAPTER 1

At the Water's Edge:
From the Aquatic Zone to the Emergent Marsh

> Magic birds were dancing in the mystic marsh. The grass swayed with
> them, and the shallow waters, and the earth fluttered under them.
> The earth was dancing with the cranes, and the low sun, and the
> wind and sky.
> —MARJORIE KINNAN RAWLINGS, *The Yearling*

Daybreak crept into the marsh slowly—hardly a sunrise, more of a smudge of gray washing across the eastern horizon. Low grunts of Virginia rails echoed through the murky morning mist. The high-pitched whinnies of sora rails reverberated, accompanied by the boink-boinking of green frogs, the slow double-toned trills of swamp sparrows, and the deep boom of a bittern. Venturing into this Iowa marsh in the predawn hours, wildlife biologist Tyler Harms felt more than a little trepidation. He was a brand-new graduate student, nervous about how his first field season would go. Although he had visited this and many other marshes like it during the day, the cacophony of night sounds made his ears ring; it was a little overwhelming and a trifle spooky. "The first time I heard it, I thought, What am I getting myself into—it sounds like there are goblins out there," Tyler recalls. But the marsh beckoned him nonetheless.

The reason for this nighttime foray was to investigate a particular group of wetland birds—the secretive marsh birds. Secretive, because they can hide in plain sight: standing motionless, their earth-toned plumages are adorned with contrasting striped feather patterns, to mimic the shadows of dark and light stripes cast by the long skinny leaves of cattails, bulrushes, and sedges. You can't see them even when you are looking straight at them. Sneaky, too, as they move quietly and fade from view at a moment's notice. This shy group of birds includes the rails—Virginia rail, king rail, sora rail, to name a few—which are all small chicken-like birds. Other cryptic and mysterious birds in this group are the American bittern and least bittern, the common gallinule, the American coot and the pied-billed grebe. Their secretive behavior and the dense vegetation of their habitat makes it difficult to discover the details of their lives: what kinds of wetlands they prefer, how they move through the day and the season, and how many of them are out there; but this is all important information for conservation. Thus, finding the answers presented a great challenge and a bit of an adventure for curious scientists like Tyler.

After donning chest waders and strapping on a heavy backpack of equipment, Tyler headed downslope toward the marsh. Gravity showed him the way to the low, roughly bowl-shaped depression where the marsh had formed, pulling him from the firm footing of the upland, into the squishy zone of fine-leaved, low-growing, grass-like sedges that grow in the low-water areas of the marsh. Because of the covert nature of his cryptic quarry, Tyler used a digital audio device to play the calls of the eight bird species he sought as he moved around through the different zones of the marsh, hoping that any birds out there would call right back. Standing among the sedges and bulrushes, he played one call, then listened. No response. Next, he played the recording of one of the other birds. Again, nothing but crickets—actual crickets, chirping. The third, fourth, and fifth species' calls also elicited no callbacks from the wild. Now Tyler was really starting to worry that his whole study would fail.

Finally, he played the recording of the Virginia rail's defense call, an ominous, loud, repetitive grunt sound. Immediately, he was rewarded: A real Virginia rail grunted right back from less than fifteen yards away. Tyler repeated the call, and the bird responded again, spot on. He was excited just to hear anything at all. Then, to his surprise, under a dawn-lightened sky, he began to see some movement in the cattails, not far from his muddy, shallow water location. As he held his breath in amazement, a chunky little bird about nine inches long ambled through the green stems and walked right up to him, stopping at his feet. Peering up at him over its long reddish bill, the rail appeared to be trying to make sense of this tall, odd-looking creature in the brown rubber suit. Lured in by a defense call, the bird was presumably expecting to meet another Virginia rail intruding on its territory. The rail tilted its head one way, then another, as if puzzling it out. Tyler stood as still as possible, holding his breath, and managed to get his camera out and take a picture—focusing straight down at the bird by his feet—without disturbing it.

Awestruck by the experience, Tyler continued to watch the dark little bird check him out. After a few minutes, the rail sauntered away, seemingly unperturbed by its alien encounter. This happened many times over the course of Tyler's two-year study, leaving him amused and amazed each time. "Those rails are pretty brave little birds. If I moved, they'd run away a little, but they would stay, watching me, checking me out. They were defending their territory, and they stuck to it as long as I played the calls. If I stopped moving, they'd come right back. Sometimes I would get two Virginia rails, both circling around me."

After that first encounter with the Virginia rail, Tyler was eager to continue the call-broadcast survey, so he waded in further, into deeper water, stopping to play the sequence of calls in the different zones of the marsh. Moving along, he could feel the bulrushes grazing his arms and the water sloshing around his legs, until he came to a stop in waist-deep water, cattails arching overhead. Once again, he played his bird-beckoning sequence of recordings, and once again, heard only crickets—and frogs—at first. But patience ever reaps its own rewards: despite the hordes of mosquitoes buzzing around his head, Tyler could hear the softest crackling of stems off to his left, and he could see the cattails moving. Slowly, slowly, a stumpy, brown-striped bird with long legs crunched its way into view. It was a least bittern, moving through the dense vegetation by clinging to the cattails. The bird eyeballed him for a few moments, keeping its distance, but slowly circling around as Tyler played the calls. Apparently concluding that Tyler was not another bittern after all, the bird then unhurriedly grasped its way out of sight. "Once they realize that you aren't a threat, they go back to what they were doing," Tyler explains. "These supposedly 'shy' birds, with the strange, tough-sounding calls, are literally tough creatures: when I moved, they wouldn't just run off like you would expect; they'd stick around to defend their turf."

Tyler spent dawn and dusk conducting this research, repeating this scenario in fifty-six wetlands across Iowa. He and his colleagues found abundant pied-billed grebes swimming in the deepwater areas beyond the cattail zones, as well as more Virginia rails, and least bitterns in wetlands with robust stands of cattails (Harms and Dinsmore 2013). It launched him into his career as a wildlife biologist for the University of Iowa. Like so many wetland researchers and managers, Tyler exemplifies a breed of scientists who are deeply devoted to the wetlands and wildlife of their home state. Proof of this dedication (er, obsession)? His ringtone is the song of the yellow-headed blackbird, and his text messages chime in with the call of the Virginia rail — two birds that find their home in the deep marsh. He's studied dragonflies and damselflies, crawfish frogs and wading birds, songbirds and dabbling ducks, as well as wetland plants and hydrology. "I've always been a wetlands person," Tyler says. "A lot of my friends call me crazy — they wonder who would want to stomp around in these habitats that are hot and buggy, wet and muddy — but I absolutely love the wetlands. They are so diverse. Everywhere you look you see something different, something new.

"After you spend enough time out in the wetland and you have these awesome experiences, you start to realize how cool these places are, and all of the difficulties of working in these habitats just fade away. You stop thinking about the one hundred million mosquitoes around your head. Instead, you focus on the damselflies and dragonflies that flush out in front of you as you walk, and on the little muskrat that swam right in front of you, heading back to its den," Tyler says. "There is so much going on in these wetlands — it is just amazing."

Most marshes, like the one Tyler studied, form in a low spot or along the shallows at the edge of a lake or river. This gradual topography creates a spectrum of water depths. First, near the top of the slope, comes the shallow marsh (or transition) zone, where the ground is consistently wet but has no standing water. Next, further downslope, is the emergent marsh (or deep marsh) zone, where the water may come up only to your ankles or all the way to your waist, as much as three feet deep. Finally, the deepest spots in the marsh form the aquatic zone, where the water depths measure three to six feet deep. Each set of water depths, or zones, harbors collections of plants that thrive in those conditions, and each set of plants supports a complementary group of insects, amphibians, birds, and mammals (table 1).

The murky water of the aquatic zone supports lily pads and submersed plants, such as coontail (*Certaphyllum demersum*) and pondweeds (*Potamogeton* spp.). In the deep marsh, the typical cattails (*Typha* spp.) and bulrushes (*Schoenoplectus* and *Scirpus* spp.) grow, edged on the deepwater side by pickerelweed (*Pontederia cordata*) and spike rush (*Eleocharis* sp.), and in the shallower spots with arrowhead (*Sagittaria* spp.) and arrow arum (*Peltandra* spp.). Grasses, sedges, and some shrubs grow in the shallow marsh zone. Of course, not every marsh has the same set of plants and animals, because it's not just the amount of water that determines what grows, but also the type of water. Whether the water in the wetland is salty or fresh, mineral-rich groundwater, silt-laden surface runoff, or pure rainwater can make a very big difference to the plants and animals that live there (see box 1).

The resulting tableau, from lily pads to cattails to sedges, nicely matches the mental picture most often conjured in people's minds when they hear the word *wetland*. It has water. It has cattails, fish, and frogs. Trees and shrubs are rare because the water is too deep (although there are some types of

Table 1. Zones in a freshwater marsh

Zone	Water depth	Plant type (with examples)	Typical wildlife
Aquatic zone	3–6 feet (0.9–1.8 meters)	Submersed aquatic vegetation (pondweed, coontail); floating leaved vegetation (duckweed, water lily)	Pied-billed grebe, mallard and other dabbling ducks, coot, heron, green frog, newt, painted turtle, bluegill
Deep-marsh zone	0–3 feet (0–0.9 meters)	Emergent vegetation (cattail, reeds, bulrush, arrowhead, pickerelweed, bur-reed, grasses)	Swamp sparrow, red-winged blackbird, rail, bittern, muskrat, mink
Shallow-marsh/ transition zone	Wet ground (water level just below-ground)	Fine-leaved vegetation (grasses, sedges), some shrubs	Marsh wren, spotted sandpiper, meadow vole, peeper, garter snake

Box 1. Water, Water, from Everywhere: Understanding Wetland Hydrology

One of the important features that wetland explorers need to understand is the way that water gets into and out of a wetland–its hydrology. By circumnavigating the wetland, poring over topographic maps, visiting several times during the year and installing monitoring wells, wetland scientists can determine where the water in the marsh comes from. The sources of water determine how deep the water or how wet the soil is and for how long. Water source also influences the types of natural chemicals and plant nutrients found in the wetland.

The depth of the water and its flow rate determine how much oxygen is available in the soil for plants, invertebrates and microbes to use (see box 2). When water is deep and stagnant, there isn't much oxygen available. If the water is flowing, or if water levels drop belowground for any part of the year, then oxygen will be able to get into the soil, which allows a number of important chemical and biological processes to take place.

Fresh water can enter a wetland from precipitation, surface water (river, streams, lakes, and stormwater runoff), and ground water. Water leaves the wetland by evaporation, transpiration (release of water vapor through plant leaves), surface water, and ground water. For wetlands near the ocean, tides bring salt water in and out, too.

Casual observers can determine a wetland's water source and learn to draw conclusions about wetland type and condition as well as the plant and animal life that lives in the wetland. All wetlands receive water from rain and snow, which bring in very few nutrients or chemicals relative to other water sources. Observers might see that the adjacent river flows up and into the marsh during high water times. Walking around the upland edge, explorers will surely see rivulets, streams, and larger channels that bring in upstream flows, or overland flood flow after a big rainfall. These sources of water tend to carry large amounts of plant nutrients most needed for growth, such as phosphorus and nitrogen.

If the water sources that flow into the wetland contain lots of nutrients, there will be more plant growth. Most marshes are quite well nourished because surface waters–rivers, lakes, and runoff–commonly flow into them. Add an abundance of sunshine and the result is an explosive growth of plant life.

The marsh may also fill up with groundwater seeping in from underneath, fed by underground aquifers. This is much harder to see, although visits in very early spring can show wet, weepy spots where the plants are greening up earlier because of the input of warm groundwater. Groundwater can contain key micronutrients, such as calcium or iron, depending on the geology of the area. Some of these micronutrients, particularly calcium and magnesium, can support unusual plant communities.

Wetlands that are fed by a lot of groundwater are called fens, and may be rich in important chemical elements. Bogs, on the other hand, receive almost all their water from rain and snow, leading to a nutrient-poor situation. See chapter 4 for more about bogs and fens.

trees that like deep water — see chap. 5). Even if many different kinds of wetlands look nothing like this one, the "classic" marsh has much to teach us. As an aquatic resource, the freshwater marsh is one of the most valuable for living creatures — both the kinds that live in it, such as marsh wrens and mallards, and the kinds that live near it, such as black bears, bobcats, and even bankers. As a biological system, the freshwater marsh harbors awe-inspiring interactions and adaptations — all hidden, awaiting discovery by a patient observer.

Life in the Aquatic Zone

How Plants Breathe

Perhaps you, like many outdoor adventurers, have guided a kayak or a canoe into the shallow edges of a lake or pond, and found your paddle entangled in coontails and pondweeds. These submersed plants signal the transition from the deep water of the lake to the aquatic zone at the edge of the marsh. As you look closely at this skein of green plant life adorning your paddle blade, you might notice that there are often two different kinds of leaves on the same plant. Pondweeds (*Potamogeton* spp.), bur-reeds (*Sparganium* spp.), and other aquatic plants often have aerial leaves that are wider and stouter than the underwater leaves, which are finely divided like very delicate ferns. This dual leaf shape, called heterophylly, is a response to the Big Problem that all wetland and aquatic plants face: a lack of oxygen (see box 2).

Oxygen doesn't diffuse easily into water, and even where the water is in direct contact with the air, oxygen diffuses only a few inches into the water column. The dissolved oxygen that is present is quickly used up by bacteria and other microbes to break down organic matter (a process called microbial respiration, essential for decomposition). Oxygen is needed for respiration, the cellular process of breaking down molecules to release energy; all cells need oxygen, even plant cells. Aquatic plants have a number of adaptations that make it easier to obtain oxygen when little is present. The differently sized leaves on aquatic plants are one such adaptation. Underwater leaves are finely divided into narrow ribbons or threads only a few cells thick, to provide more surface area for the limited amount of oxygen to pass from the water directly into each cell. On the very same plant, the leaves that lay on

---

Box 2. Surviving the Flood: Plant Adaptations to Standing in Water

Author Sharon writes: I water my plant until the water fills the dish below the pot and think that ought to take care of it for a while. When the water is gone from the dish, I repeat the exercise because it's easier than sticking my finger in the dirt every day to judge moisture, and in the long run I'll have to water the plant less frequently—I just gave it an extra supply, after all. After a few days, the plant turns yellow and wilts. Huh, the poor thing must be thirsty. So I water it again; it dies. "But I was taking care of it," I whine to my mother. "I watered it!" Flooded it, to be precise. And maybe this is why I love wetland plants: they can handle what my poor philodendron could not.

While I'm still far too irresponsible to be trusted with all but the hardiest house-plants, I now understand why the philodendrons, the dieffenbachias, and the spider plants die. As I fill the soil pores with water, the air is pushed out and, with it, readily available oxygen. Plants don't only produce oxygen; they consume oxygen just as animals do for respiration, the process of breaking down carbohydrate molecules and using the resulting energy for growth and reproduction. If I keep the soil saturated, the plant rapidly uses any available atmospheric oxygen. Oxygen can still diffuse from the atmosphere into the water now filling the soil pores, but it moves a lot slower through water than air—too slow for the needs of the plant. In response, the plant's metabolism slows and water uptake declines. Ironically, the plant attempts to conserve water by closing its stomata (the pores in the leaves), just as it does under drought conditions. Photosynthetic activity declines and the plant's cells become flaccid, no longer plump with water pressing against cell walls, and the plant wilts just as it would if I had not watered at all. To continue functioning, the plant shifts to anaerobic (no-oxygen) processes, which yield far less energy for plant maintenance, produce the byproduct ethanol, and acidify the cell environment—all of which creates a rather unhappy situation for your plant, regardless of how encouragingly you talk to it. Additionally, anaerobic conditions change the soil's chemistry, converting (reducing) minerals like iron and manganese to toxic forms, which are lethal to plants not adapted for such conditions. My poor houseplant didn't stand a chance.

But where there is a niche, there is a way; and wetland plants have numerous adaptations for making the most of a stressful situation. Hydrophytes (from the Greek *hydro* for "water" and *phyton* for "plant") adopt a number of strategies to ameliorate a low-oxygen environment: the plant can obtain oxygen from somewhere other than the saturated root zone, the plant can neutralize the toxicity of reduced minerals in the soil, it can change its shape to maximize oxygen intake, or it can attempt to "hold out" during flooding. Many wetland plants multitask, using one or more morphological or physiological game plans, and thus survive where upland plants cannot.

A common strategy is the formation of porous plant tissue, called aerenchyma, which allows diffusion of oxygen from high concentrations in the aerial parts of the plant to lower concentrations of oxygen in the roots. Aerenchyma tissue, which essentially creates "air pipes," forms as cell walls disintegrate or move apart from one another as part of normal plant growth or when triggered by a lack of oxygen. While any houseplant will have some aerenchyma tissue, maybe up to 10% or so of the plant—wetland plants may be up to 60% aerenchyma tissue.

Peel a cattail leaf lengthwise and notice the honeycombed spaces in the leaf, or cut a cross section of a pond lily stem to see the large open air pockets; oxygen can diffuse through these spaces relatively quickly to submerged roots. Of course, those lilies growing from underwater rootstock must get some portion of their anatomy to the surface before drowning, so they are able to rapidly elongate stems to reach the surface and then fold out those beautiful floating leaves to breathe and soak in the sun.

If enough oxygen is transported down from aerial plant parts to inundated root tissues, it leaks from the roots into the surrounding soil, detoxifying the minerals through a chemical process called oxidation and creating what is called an oxidized rhizosphere—an oxygenated zone of protection for plant roots. If you pull out of the soil an arrowhead plant (*Sagittaria* spp.), some cordgrass (*Spartina alterniflora*), or one of the wetland sedges (*Carex* spp.), you may see traces of a rusty orange color along the roots, an indication of oxidized iron.

Plants not inundated for long periods of time may survive seasonal or temporary flooding by storing carbohydrates much like we might preserve food for lean times or a power outage. Production of ATP—adenosine triphosphate, the energy molecule built during photosynthesis—slows dramatically under anaerobic conditions so plants with thick, starchy rhizomes can withstand flooding longer, living off their stores of carbs. The highly invasive *Phragmites australis* (common reedgrass) employs this strategy, much to the annoyance of wetland managers trying to rid East Coast marshes of this towering, dense grass.

And some plants adapted to flooding just keep their "feet" up. Adventitious roots grow from the portion of the stem above the soil and thus are bathed in atmospheric oxygen while deeper roots cease to function. You may have seen such roots on waterlogged crop plants that utilize this strategy, but only for a short time before succumbing to the inevitable paucity of oxygen belowground.

Trees in floodplains typically have shallow root systems, keeping vital rootstock above the water table, closer to the oxygenated part of the soil. Red mangrove swamps are dense forests of woven and tangled prop roots holding root tissue, stems and leaves above the waterline. Black mangroves produce pneumatophores—roots much like rotten fingers reaching up through the water to capture air.

(continued)

Plant accommodations for sodden conditions are varied and involve a complicated sequence of events driven by hormones and chemical signals that result in structural and physiological changes to the plant. So, without such adjustments, how do land plants grown hydroponically survive? Hydroponic and hydroculture plants are not stressed by lack of oxygen. Your hydroponic tomatoes are grown in water that is aerated with a pump system, and most of the root mass of a hydroculture houseplant is kept out of the water.

Despite understanding the nature of overwatering, or underwatering, my houseplants, I am just too erratic and they are much too finicky about soil moisture. Certainly, somewhere between marsh and desert, there is a plant out there I can keep alive in my home. If not, I'll have to content myself with growing backyard lettuce and tomatoes, but that is another touchy subject.

---

the surface or stick out of the water will have a different outline — maybe like a paddle, an arrow, or a three-lobed clover. These surface leaves, which are in contact with the air, have far less of an oxygen problem, so they are wider in order to maximize area for photosynthesis. They also must be thicker to support themselves out of water.

Even those plants with floating leaves have to deal with the oxygen problem resulting from most of the plant being underwater. The large floating leaves of water lilies — big and heart-shaped for the yellow water lily (*Nuphar lutea*), sharply cut lobes for the white water lily (*Nymphaea odorata*) — are surprisingly dry. Leathery textures allow water to roll off quickly, permitting the large numbers of pores on leaf surfaces to bring in more oxygen, which is pumped to plant roots in the muck below.

Gliding through this flotilla of water lilies in your kayak, you may spy another feature that helps the water lily adapt to low-oxygen conditions. A long, thick, dark-brown, scaly-looking entity, a foot or more long and several inches wide, may appear at first glance to be some kind of reptile. However, this primitive-looking structure is actually a water lily rhizome that has floated up from the mucky bottom. Rhizomes are a type of underground stem that can produce both roots and shoots, and which store lots of carbohydrate-rich food, a product of photosynthesis. In order to convert carbohydrates back into energy for growth, the plants need oxygen to reach their underground parts on the muddy bottom, and these weird-looking rhizomes play a key role as an oxygen pump. In the rhizome, pressure from younger airborne leaves pushes through the leaf stems and into the rhizomes and roots, and out through the stems of the older leaves. This oxygen pump develops when warm temperatures create higher humidity inside the plant than outside it. More humidity means more water molecules inside the leaf cell, and fewer oxygen molecules. This causes oxygen to move from outside the plant, where there is a higher concentration of oxygen in the air, into the leaf, where oxygen concentration is lower. More oxygen inside the leaf creates higher gas pressure. Younger leaves have smaller pores on the outside, supporting these higher pressures. Young leaves are also more likely to be red-tinged, which warms the leaf faster and builds more pressure. Older, larger leaves have larger pores, which don't hold the pressure and thus allow air to escape. Escaping air creates airflow from the young, highly pressurized leaves, through the stems, down to the roots, and up again through the stems of the old, leaky, low pressure leaves. The pressure gradient brings more oxygen into the plant, helping it survive. Similar kinds of pressure-induced airflow also take place in many other wetland species (Willey 2016; Cronk and Fennessy 2001), creating an underwater jungle in the aquatic zone of the freshwater marsh.

Rolling in the Deep: Insects in the Aquatic Zone

On a warm day in May, a few counties away from the Iowa marsh where Tyler Harms found his secretive marsh birds, two hundred middle school kids disembark from their bright yellow buses and career downslope to the backwater marshes of the Mississippi River in New Albin, Iowa, running, laughing, shoving, shouting. To a hapless bystander, the scene appears to be something between a chaotic picnic and a jailbreak; to the students, it is both. To Jackie Gautsch, biologist with the Iowa Department of Natural Resources, it is just another day immersing the next generation in wetland ecology—literally and figuratively. The marshes they set out to explore are the backwater sloughs of the Upper Mississippi River National Wildlife and Fish Refuge, a set of marshes still connected to the river and thus very diverse. "Most of them are farm kids. They spend a lot of time outside," Jackie says, "but this isn't something they do in their free time, so they are just fascinated by all the creatures they find." She points out that 90% of Iowa's wetlands have been drained. The hope is that this experience will set the young people up to understand the importance of these ecosystems, and to become explorers for the rest of their lives. Their goal today is to find as many invertebrate animals—creatures without backbones, such as insects and worms—as they can, and use this information to evaluate water quality.

In groups of fifteen at a time, the students pull on waders, grab equipment, and head into the marsh. First, they wade through the sedges and grasses at the wetland's edge, then they move into the deep marsh, encountering dense stems of cattails and bulrushes. "This is the age group that likes to test limits! We tell them not to go beyond their knees and to stay in the emergent marsh, but they head right out to water as deep as their waists to the submersed vegetation because that's where all the cool stuff is, out with the aquatic weeds and lily pads, and it is something they have never done before," Jackie says, anticipating the splashes and falls to come.

Walking into the marsh, all the middle schoolers quickly notice that the water gets deeper as they walk further from the upland. Realizing it's easier to move through the open-water aquatic zone than the dense cattails, our intrepid marsh explorers splash right into the deeper areas, using dip nets to sweep through the water to capture their quarry. Abundant plant life provides food and hiding places for a large variety of swimming bugs, diving beetles, delicate mayfly nymphs, armored juvenile stoneflies, worm-like mosquito larva, creeping crawdads, and tiny crustaceans. Some of these wee beasties shred the dead leaves that fall into the marsh into bite-size chunks of nourishment; others chow down on the tasty greens of living plants; and still others hunt their fellow invertebrates for a meal. In turn, this vast cast of characters becomes a key food source for larger animals—notably, the fish, amphibians, and reptiles who also call the marsh their home.

Sorting through their catch, the students identify different invertebrates and classify each one as pollution tolerant or intolerant. Backswimmers, pill bugs, tiny shrimp-like "scuds," and prehistoric-looking dragonfly larva are all part of the catch. But the one they will never forget is the water scorpion (*Ranatra* spp.). "It looks like a walking stick—it has a long tail-like breathing tube on its back end, which it uses like a snorkel to get air from the surface," Jackie explains. "Water scorpions are predators, so they have large forearms sticking up, and they use their piercing mouthparts to stab their prey, liquefy their insides, and suck out their juices. The kids love that story, and they even let the water scorpions walk on their hands."

By counting the number of pollution-tolerant and -intolerant inverte-brates, the students can determine if there is a water-quality problem in the adjacent section of the river that overflows into this backwater wetland swale they are exploring. They learn that without clean water, and without wetland habitat, these aquatic creatures will not survive; nor will the fish, reptiles, and birds that feed on them. And while taking all this in, Jackie notes, "they all have a good time. There are always a couple of kids that lag behind — they are the last ones that leave the session; they hang around and they want to identify every single bug they find." Future wetland scientists, perhaps? They've already realized that exploring a wetland takes patience, keen observation skills, good hand-to-eye coordination, and a wicked good sense of adventure.

Returning to this same spot later in the season, these students would see that the leaves of water lilies look decidedly beaten up — riddled with patterns of twisting, curving trenches. Not only do aquatic plants have to adapt to the low oxygen levels, but, like all plants, they have predators. The trenches on the water lily leaves are the feeding paths of the brown or some-times glittery-gold water lily leaf beetles (*Galerucella nymphaeae* as well as species of *Donacia*). Although water lilies, as well as other plants, produce chemical defenses to discourage this munching, these beetles have evolved to tolerate the poisons. Once chewed, a grazed leaf is more susceptible to bacterial attack, so the leaf decays more rapidly, then drops to the bottom. Where there are a lot of these beetles, the leaves may last only about seven-teen days (Wallace and O'Hop 1985). To compensate, plants that are grazed by beetles will grow one and a half times more quickly, producing leaves as fast as the beetles and bacteria destroy them (Cherry and Gough 2009). This is possible only where the marsh mud is high in nutrients needed to support rapid plant growth. Though the beetles may seem destructive, they are important nutrient recyclers, and in turn they find themselves served up as excellent snacks for passing mallards, rails, and other birds, as well as fish and amphibians.

But the plants here in the aquatic zone are sometimes the predators, too.

While examining the Gordian knot of plant matter adorning their nets, our group of student bug hunters may also notice dozens of tiny oval pods protruding at regular intervals from a mesh of very thin, grayish-green threads. These pods are in fact the leaf bladders of the aptly named bladderwort (*Utricularia* spp.), a carnivorous plant. Each tiny bladder sets a trap by first expelling water. When a small, tasty creature such as a tiny copepod swims by, minute hairs near the bladder's opening are disturbed, triggering the bladder to open. The copepod is sucked into the bladder in less than a millisecond as the bladder fills with water. Within an hour, bacteria and enzymes inside the bladder have digested the little animal, providing much-needed food for this sneaky plant, which has no roots and thus needs to obtain nitrogen, phosphorus, and other nutrients from its prey.

Four Legs in Deep Water: An Amphibious Life

Of all the creatures in the deepwater aquatic zone, nothing captivates young people more than trying to capture the multitudes of amphibians swimming about. Unlike reptiles, amphibians need to stay moist, and they must all lay their eggs in water or in very damp spots. So, for at least part of their life cycle, frogs and salamanders will be found in an aquatic setting. As summer comes on, warm temperatures bring the slow bonk, bonk of the green frog (*Lithobates clamitans*), its call resembling the plucking of a loose banjo string; the long staccato trill of the American toad (*Bufo americanus*), common in much of the northeastern United States; the squeaky-door-hinge call of the pickerel frog (*Rana palustris*), and the deafening bass of the bullfrog (*Lithobates catesbeianus*). After mating, each species releases masses of eggs in long strings, loose jelly-like masses, or tight clusters in the murky water of the pool area. Hatching tadpoles will hide under leaves in the sun-warmed water at the pond's edges. Adult frogs will sunbathe, hopping into the mucky areas just out of the water. Walking along the pond's edge often seems to set off a cavalcade of frogs retreating into the water and quickly swimming off a short distance, only to turn and watch the intruder's passing.

The open-water areas of the aquatic zone offer great space not only for frogs, but also for an interesting little hunter: the eastern spotted newt (*Notophthalmus viridescens*). "Newts are pretty uncommon in the wetlands of

Iowa," Jackie Gautsch explains, "so the kids had a ball trying to catch them." The greenish-brown, fin-tailed newts may be as small as three inches—hardly longer than a human thumb—or almost as long as a whole hand. Wading around the aquatic zone at the edge of a pond, visitors can see newts hanging motionless, their light-colored underside blending in with the sky if seen from below, while their darker backs camouflage against the murky bottom when observed from above. Newts often seem to be positioned at least two dozen body lengths from one another, in temporary hunting territories, where they watch for unwary swimming beetle grubs, water mites, worms, leeches, and small shrimp-like freshwater crustaceans. These cagey little predators are even significant consumers of mosquito larvae, with each adult eating up to one thousand baby mosquitoes a day (DuRant and Hopkins 2008). Hail the newts!

The breeding habits of eastern spotted newts are also intriguing. In shallow areas at the edge of a stream, lake, or reservoir, thousands of newts collect each spring in a swarming mass, performing undulating hula dances to attract mates. Each female lays two hundred to four hundred eggs in clusters attached to submerged vegetation or falling leaves; most of these eggs are eaten by predaceous aquatic larvae of beetles, dragonflies, and other insects. After as many as fifty days, the eggs that survive produce fingernail-size tadpoles with feathery collars for gills.

Newts have a strange life cycle that sets them apart from other salamanders. By the end of the summer, each tadpole transforms into a small bright orange eft, the juvenile stage of the newt (and an excellent word to know for the game of Scrabble). The efts are terrestrial, their flame-orange color warning potential predators that toxic chemicals in their skin would make for an unsatisfying dining experience. After spending about two years wandering the woods, the efts turn a darker brown and develop a fish-like tail shape. They then return to their natal pond, to hang out and seek revenge on the water-dwelling bugs that long ago ate their potential brothers and sisters. In some parts of the country where permanent flooding is the norm, a few subspecies of the newt forgo the eft stage completely, and the tadpole morphs directly into either a form of the adult with gills or one with lungs (Takahashi and Parris 2008). By skipping the wandering teenager phase of life, these purely aquatic newts are protected from the dangers of the forest;

but one wonders if, in doing so, they will end up with fewer good stories to tell their grandchildren.

Fish Stories and Plant Jungles of the Aquatic Zone

Deepwater aquatic zones along the edges of large streams, rivers, and lakes provide critical habitat not only for frogs and newts, but also for fish. Across the river from our middle schoolers' explorations, on the Wisconsin side of the Upper Mississippi River, biologist Jeff Janvrin has spent several years trying to discover how the fish use the wetlands along the river—areas of submersed aquatic vegetation, emergent marshes and muddy-bottomed ponds called sloughs.

As a habitat specialist with the Wisconsin Department of Natural Resources (WDNR), Jeff is interested in understanding how the diversity and numbers of fish have changed over the past century since the series of locks and dams were installed in the river for shipping and flood control. Behind the dams, the water is much deeper than it used to be, the flow patterns in the river have changed, and many of the low-lying islands have been flooded or worn away by the currents. "Mark Twain's Mississippi River is not the Upper Mississippi River," Jeff says. "Twain described the lower Mississippi, where the river meandered from bluff to bluff," before it was channelized and dammed In its northern sections, the river didn't meander as much. "There are islands where archaeologists have found spearheads that are ten thousand years old, indicating that these backwaters, away from the main channel, have been stable for a long time. There were lots of floodplain forests, intermixed with shallow ponds and grassy marshes. Today, the lock-and-dam system has changed that structure, and we want to know how that affects the fish."

Bluegills and the largemouth bass are the fish locals are most keen to understand. "People don't realize that fish migrate, just like birds do. Largemouth bass will move up to fifteen miles between their summer and winter habitats," Jeff explains. Come spring spawning season, bass and bluegills build their nests in a lot of different areas, quiet shallow marshes and backwaters that don't have too much flow. In summer, they want less vegetated areas—the sloughs and channels—where the oxygen levels are stable. For successful overwintering, the bluegills and largemouth bass need deeper, calmer waters, which are usually found in the backwater ponds. Before the river was dammed, there were lots of sloughs for overwintering. After the dam, the sloughs had too much current; the only quiet water was found in the upper ends of the sloughs. Because overwintering habitat can be a limiting factor for the fish population, the WDNR needed an inventory of the areas that provide this sweet spot of just-right water depths. To find out, Jeff and his team employed some shocking techniques—electroshocking, to be specific.

A two-person crew, outfitted in a special electroshocking boat, makes the annual rounds of the sloughs in the fall of each year. Jeff describes the scene: "On cold, overcast days, we wear survival gear. But I can remember brisk fall days of glass-calm water, under clear blue October skies. Around us are the huge beautiful river bluffs with all the fall colors." From the boat, they send an electric shock into the water, which stuns the fish temporarily. While the driver controls the boat, the other person scoops the stunned fish into the holding tanks. Then, with the boat anchored, they count and measure the fish. Exploring all these aquatic zones and deep-marsh areas over a seven-year period, Jeff and his colleagues discovered, not surprisingly, that most of the overwintering habitat is in the upper sections of the pools, well upstream from the dams, where the water is shallower and current velocities are slower. There are about one-third fewer overwintering sites than there had been before the dams were built.

"It's really a vast wilderness area out here along the Upper Mississippi," Jeff notes. "It's a huge wetland. People know about its importance for migratory birds, but since fish are hard to see, people forget about them. They don't realize that we have one hundred and forty fish species; it is one of the most diverse fish habitats in the US. And it's not just one thing that maintains healthy populations of fish. It's all interrelated, and we have to keep all of the resources they need, and we need to keep all those pieces connected. The restoration projects we have done show what we can do when we work together, and when we keep a broad perspective on all the creatures that live here."

Gretchen Benjamin, who works for the Nature Conservancy in Wisconsin, describes one such wetland restoration project on the river. Earlier in her career, while an employee with the WDNR, Gretchen worked with a large group of stakeholders to explore ways to restore the wetland and aquatic zones that had been drowned out by the dams. "The aquatic plants are the basis of the ecosystem—we had to get them back in order to have healthy fisheries and waterfowl and mammals and invertebrates. So, we asked the question: Can we use the dams to reduce the water levels in the summer months, in order to recreate the conditions that favor the aquatic plants?"

This proposal was not well received at first—people thought it would affect their ability to enjoy the river. "The river is a way of life around here; there are people who have spent fifty to sixty years out on the river. But the river people—self-described 'river rats'—knew that the river needed some help. They noticed that when the aquatic plants went away, the fishing wasn't quite as good, the duck hunting wasn't quite as good."

In 2001 and 2002, they used the dams to drop the water levels by eighteen inches in a twenty-eight-thousand-acre area called Pool 8 near La Crosse, Wisconsin. Shallow muddy areas that had not been out of the water in seventy to eighty years immediately started to green up with sedges, pigweeds, and arrowheads. "The response was phenomenal in the backwater areas as well as the border of the main channel, which had plants growing where no plants grew before. We did it the following year, and the plants really became strong and robust," Gretchen explains. This tactic of decreasing water levels behind the dams to recreate natural conditions has been replicated in other places since then, and teams of scientists are working to do this on a more routine basis to establish thousands of acres of aquatic plants. "When I first saw it happening, it brought a tear to my eye. As for the locals, when they saw the plants showing up, they started seeing more birds and catching more fish, and they realized that the aquatic plants are essential parts of the ecosystem. This restoration allowed us to put the basic elements back together again."

Life in the Emergent Marsh Zone

Still the Problem of Oxygen

Upslope from the open-water aquatic zone, many plants poke up through as much as two feet of water—these are called "emergents." Their roots reach into the bottom sediments, creating an unimaginably rich tangle belowground. The cattails, bur-reeds, and arrowhead plants found here all employ a number of adaptations to deal with the stress of sitting in water for most if not all of the year. As explained in box 2, their evolutionary response is to develop "air pipes" (called aerenchyma) in their stems, which allow oxygen to diffuse into the roots. In plants such as bur-reed, pickerelweed, and arrow arum, these pipes, called aerenchyma, give the stems a crispy but spongy feel when squished between thumb and forefinger. In many species, if you compare the leaves from two different areas by cutting them with a knife to see the cross section, you will see more aerenchyma in the leaf that grows in the wetter area.

Once they evolved tricks to pump oxygen down to their roots, these spongy-stemmed emergent plants were able to take advantage of the fine growing conditions of the marsh. Open light, plenty of water, and abundant nutrients add up to fast, strong, and impressively dense growth, as anyone who has ever tried to walk through a marsh knows all too well. The most common emergent plants, such as broad-leaved cattails (*Typha latifolia*), are generalists which can withstand a very wide range of water levels, from barely saturated to three feet deep (Magee and Kentula 2005). Their large rhizomes can persist without oxygen for up to four months; the root uses anaerobic pathways of respiration to get energy, which means the cells don't need oxygen to get usable energy out of the stored starch made during photosynthesis.

Some emergent plants employ other specialized adaptations to bring in and hold more oxygen. One of these is the common reed *Phragmites australis* subsp. *australis*, an often-invasive plant from Europe regularly seen growing in dense clumps in wetlands and roadside ditches. With its very tall (sometimes more than twelve feet!) green stems, topped with fuzzy tan brooms of seeds, the common reed is hard to miss: once you recognize it, you will start seeing it everywhere. It is usually an aggressive plant that takes over when nutrient levels rise with incoming stormwater pollution, water levels change, or the plants simply moves in from an adjacent wetland. The common reed has hollow stems, which bring in more oxygen through pressurized gas flow. Similar to the water lily, air inside the plant is more humid than the air outside, so gas (oxygen) flows into the plant through pores on the stem called stomata, building up pressure. The air then flows out through any broken, standing-dead stems, which act as high-pressure "exhaust" pipes. Bringing in more oxygen allows the reed's underground parts to grow as fast as four inches a day, helping it colonize and take over wetland habitats. The plant's belowground roots and rhizomes are so thick that an exposed section looks like a densely woven basket.

Thanks to these and other adaptations, the nonnative form of the common reed is highly invasive, particularly on the edges of salt marshes, where fresh water flows in from upland borders (see chap. 7). It spreads by seed and by underground roots and rhizomes, taking over the territory formerly occupied by sedges, cattails and other native plants. While the common reed does provide habitat for some insects and wildlife, these dense stands don't support the same level of biodiversity as areas where the reed grows in lower densities.

But there is an interesting twist to this story. As some people had long suspected, there are several different subspecies of *Phragmites australis*, and a few of them are native, rare, and not aggressive (Allen et al. 2017). Dave Burdick, research professor at the University of New Hampshire, had been studying the pernicious invasive type for a long time when he became intrigued by the idea that there was a genetically distinct native variety. "We knew that this species had been around a long time, because researchers have found roots buried in sediments three meters deep from three thousand years ago," he says. The native type of common reed [*Phragmites australis* subsp. *Americanus*] looks a little different, showing a chestnut red tinge on the stem.

With thoughts of the native reed in the back of his mind, Dave visited Sandy Point marsh in Great Bay, New Hampshire, with naturalist Liz Duff on a dreary November day to examine an area where the New Hampshire Natural Resources Conservation Service had been trying to control the aggressive form of common reed. "We were walking around, and then we both looked up at the same time and saw this pink-and-white stripe on the stems—'Oh my gosh, this is the native plant!' We took samples to an expert and confirmed that we have the largest—and maybe the only—stand of native common reed in the state of New Hampshire. Here we were trying to kill it, because it looks so similar to the aggressive form, and it turns out to be perhaps one of the rarest plants in the state, hiding in plain sight!"

This completely changed Dave's perception of the area; he no longer saw it as a stand of an unsightly aggressive plant, but as a unique area harboring a rare plant and a healthy natural community. "A diverse group of plants grows with the native common reed because it doesn't grow so densely. Nature is there, giving us the answers, if we care to look or know how to look. All of these aha moments come from using our observational skills to look at nature in a slightly different way."

Birds of the Emergent Marsh Zone: Ambassadors, Dabblers, and Divers

Fabulously thick jungles of emergent cattails, grasses, and reeds produce enormous amounts of food for the invertebrates that live there, as well as for the waters downstream which receive the bounty of exported stems, leaves, insects, and other materials carried by the outflowing currents. The tangle of plants also creates excellent nooks and crannies for animals to nest, rest, and hide.

Those who benefit directly from the food and shelter include one of the rarest flyers: the whooping crane. "It's a large white bird, you can't miss it!" exclaims Brenda Kelly, biologist for the Wisconsin Department of Natural Resources. Although an avid birder, Brenda had never seen one of these highly endangered birds in the wild. In the early 2000s, when whooping cranes returned to the wetlands of Wisconsin from releases of captive-reared birds and their progeny, birders were on the lookout for these majestic animals. While zooming around the five-thousand-acre Mud Lake Wildlife Area in Columbia County, Wisconsin, in an airboat, Brenda and her coworker came around a corner in a narrow cut and came face-to-face with two whooping cranes. Standing almost five feet tall at the edge of a cattail marsh, probing the pondweeds and mud for roots, snails and tadpoles to eat, they took no notice of the noisy intruders. "We shut off the airboat and just floated by; it was perfectly quiet. It was just beautiful." she marvels. Since then, visiting the marsh to see the whooping cranes has become a major tourist attraction — for the intrepid tourist willing to walk a good distance on the dirt roads bisecting the marsh. "People love seeing the whooping cranes. They go out into that wetland, and seeing the cranes leaves a mark on them. The whooping cranes are the greatest ambassadors for the wetlands, bringing new people out to get connected to the marsh, and showing them that these wetlands are special. The fact that these birds chose this wetland shows that the marsh is a vital and important habitat."

Many of the marshes of the midwestern and northeastern United States were formed by huge chunks of glacial ice, scattered behind the retreating glaciers, which left bowl-shaped depressions in the mix of clay, sand, gravel, and rocks as they melted. This rocky material, also known as glacial till, was deposited in uneven, bumpy fashion, sometimes resulting in an assortment of small ponds strewn across the landscape. In northern Iowa up through western Minnesota and the Dakotas, Saskatchewan, and Alberta, these networks of ponds are called prairie potholes. Ranging in size from quarter-acre patches to shallow lakes as big as a farm field, they play a critical role for migrating and breeding birds—particularly, waterfowl and shorebirds along the bird migration route known as the central flyway. Paul Errington, a duck hunter and wetland writer in the mid-1900s, describes the rich variety of life found in these vital habitats:

> No one view could typify a marsh of eastern South Dakota at its life-rich summer best. One view should be of a misty morning with sunlight filtering through, and avocets, willets, and lesser shore birds running along a mudflat, feeding, raising their wings . . . and calling. On mud- or sandbars or floating posts or muskrat lodges, the terns guard their territories. Over all, the medley of blackbird and bobolink calls, of coot and rail and grebe calls, the pumping of bitterns. In the right places, the booming of prairie chickens is part of the morning sounds of early summer. Ducks are much in the marsh picture. . . . On shore are the mallards and pintails, the baldpates, shovelers, gadwalls, green-winged teal, and especially the blue-winged teal, bluewings everywhere. (Errington 1957)

Ducks nest in the dense cattails and on the muskrat lodges. Diving ducks swim in the murky water, gobbling down shiners, frogs, aquatic insects, snails and crayfish and clams. To avoid competition, different duck species specialize in different parts of the marsh: the canvasbacks fish from the bottom of the open-water area, while redheads feed in the shallower areas near the edge. The dabbling ducks, such as mallards, teal, and shovelers, dip their heads down, waggling their rears as they pull up nutritious arrowhead tubers (also known as duck potatoes), cattail rhizomes, seeds, and aquatic insects (Ehrlich, Dobkin, and Wheye 1988).

Holding water in dry, treeless landscapes, these prairie potholes are keystones for the birds that stop by to refuel themselves as they migrate from South America to Alaska and back again. These potholes are also critical for humans, recharging the groundwater by holding on to scarce surface water and slowly releasing it into the underlying aquifers, even during prolonged droughts (Winter and Rosenberry 1998). Sadly, many pothole marshes have been drained and planted to crops, contributing to both the decline in bird populations and water table levels.

Of Marshes and Muskrats

As the saying goes, the only constant is change, as true in a wetland as anywhere else: plants and animals come and go, water levels change, nothing ever stays the same. These changes can originate from within the marsh, or in the watershed beyond it. The outside world changes—a road is built, a river meanders, a forest becomes a farm, the earth's climate is altered—and the water coming into and leaving the wetland changes. Maybe more rainfall, more stormwater, bringing more farm fertilizers. More pavement in the watershed means less groundwater flows into the wetland, which means less sweet-soil calcium seeping into the marsh mud. In addition to these external factors, internal agents of change—the plants and animals themselves—also profoundly affect the marsh. The cattails grow more densely, the sedges start forming tussocks, the water levels drop as plant roots turn liquid water into water vapor.

Freshwater marshes were once thought to be in a constant state of change in a particular direction. With lots of sunshine, water, and nutrients, aquatic plants at the edge of a pond use the miracle of photosynthesis to convert the carbon dioxide in the air into leaves, stems, flowers, and roots. Every fall, much of this growth is then dropped into the water below. These lush pieces of organic matter feed armies of bacteria in a dry setting, but underwater, there isn't enough oxygen for the decomposers to do their duty. Rot doesn't set in. Very little of the dead stuff decays; it accumulates year after year. The bottom of the pond fills up, and the water becomes shallower, allowing marsh plants to move in. The process repeats: grow, drop, fill up.

Taking this process to its logical conclusion, people once thought that eventually every lake would fill in and become a marsh, and every marsh would dry up and turn into an upland forest or prairie. After much study, however, scientists realized that this scenario doesn't actually hold water, because the dead materials would stop building up once they reached the surface of the water and were exposed to air. Rot *would* set in then, and the building process would stop. Studies of the pollen record show that some lakes did become dry terrestrial forests, but it was the result of external forces—natural climate change—not the buildup of organic matter past the water level. Still, shallow ponds do fill in, very slowly in most cases, and a forested wetland can result. It just won't get past the wetland stage (Mitsch and Gosselink 2000). And many, many events can redirect this process onto a different course.

Muskrats (*Ondatra zibethica*), for example. As a marsh full of emergent plants gets to its most densely packed state, these little three-pound rodents react like hungry homesteaders when a new diner opens up: they flock in from far away, so pleased to have a new restaurant. On the menu are the carbohydrate-rich roots, shoots, and leaves of cattails, bulrushes, sedges, arrowheads, water lilies, and pondweeds (Snyder 1993). They will also grab an occasional high-protein snack of tadpoles, crayfish, or freshwater mussels, both the endangered kind (e.g., dwarf wedge mussels) and the invasive kind (e.g., zebra mussels).

Once they find this great food source, muskrats tend to stay on, producing as many as three litters per year. They build small lodges, two-foot-high mounds of mud, leaves, and sticks, as well as raised beds of comfy materials to serve as feeding platforms, keeping them dry while they dine away from home. These little rat decks are sometimes even equipped with nice pergolas, for shade and rain protection. Muskrats do know how to live.

As they move around, these swamp rats trample the vegetation into a network of trails. By gathering food as close to home as they can, the area around their lodges is quickly made devoid of plant life. Before long, what was an impenetrable fortress of cattails becomes a series of jungle islands, surrounded by moats and bisected by trails. The result is more open water and more sunlight, bringing in a greater number and more diverse plants—not just the monoculture of two or three dominant plants. Among the plants that will be found in greater abundance thanks to muskrats are the delicate, water hemlock (*Cicuta bulbifera*) with its ferny leaves, the red-tinged leaves and stems of marsh Saint-John's-wort (*Triadenum virginicum*), and the strangely long and skinny flower buds of marsh willow herb (*Epilobium leptophyllum*) (Hewitt and Miyanishi 1997). Thus, the muskrats' quest for nourishment creates new habitat for other plants, which in turn support different kinds of insects and birds — wetlands in transition. The lodges muskrats build also support numerous other animal species: geese, swans, teal, canvasbacks, and pied-billed grebes will nest on top; stinkpots, snappers, and other turtles will hibernate and lay eggs inside; water snakes and painted turtles will bask on the exterior (Kiviat 1978).

Tides of Fresh Water in the Marsh

Muskrats are common critters in almost any cattail marsh. They are particularly happy in the freshwater marshes found on the upper reaches of estuaries — bodies of water that feel the pull of the sea. So much fresh water flows into these areas from upstream that any salt water from the ocean is highly diluted, but they are still influenced by the tides. While muskrats are still important drivers of diversity in the tidal freshwater wetlands, many of these marshes have a larger variety of plant species than freshwater or saltwater wetlands elsewhere, even without the muskrat's trail-tromping and cattail-chomping influences.

With no harsh salt water to contend with, but with abundant sediment and nutrients from upstream waters and tide-driven bottom flows from downstream, the tidal freshwater marsh enjoys high plant productivity and diversity in the upper portions of the wetland (Barendregt and Swarth 2013). For example, along the York River, which flows into Chesapeake Bay, there is a transition from freshwater marshes upstream to salt marsh further downriver. Rob Atkinson (whom we met in the introduction, stuck in the tidal Sweet Hall Marsh along the Pamunkey River, a tributary of the York River) and Jim Perry found fifty-six different plant species in the freshwater tidal marsh, and fewer than twenty species in the saltier wetlands (Perry and Atkinson 1997). The tidal freshwater marsh often looks markedly different than its nontidal freshwater wetland cousins as well: cattails may not grow quite as densely here, instead sharing their space with the arrow-shaped leaves of arrow arum (*Peltandra virginica*), along with the juicy and delicate stems of jewelweed (*Impatiens capensis*), the pink or white pendulant flower clusters of smartweeds (*Polygonum* spp.), and the showy pink rose mallows (*Hibiscus moscheutos*). Healthy stands of wild rice (*Zizania* spp.), cattail, and giant reed are also common. On the flowing edge of the marsh, undulating beds of green arrow arum and purple-flowered pickerelweed (*Pontederia cordata*) grow in abundance, bordered on the river side by yellow spatterdock (*Nuphar* spp.), eelgrass (*Zostera marina*), and waterweed (*Elodea* spp.).

During high tide, the higher parts of the marsh, near the uplands, may be covered with water for two or three hours, while the soil of the lower sections, near the river, may only be exposed for brief periods during low tide. Exposing the soil allows oxygen to flow into soil pores and plant roots, thus preventing many of the problems associated with stagnant, saturated conditions (Whigham and Simpson 1992) (see box 2, above). Some high-quality tidal freshwater marshes have been found to harbor from 60 to 137 plant species (Perry et al. 2009). Of course, not all tidal freshwater wetlands experience such perfect conditions, as many have been subjected to heavy pollutant loads, altered hydrology, and invasive species.

In addition to these daily tidal changes, there are many seasonal changes. Botanist Mary Allessio Leck, professor emerita of Rider University in New Jersey, describes how the tidal freshwater marsh changes through the seasons. At the end of winter, it is drab, full of every hue of brown—from the dark, bare brown mud along the river edges and the varied browns of the dead plants, to tawny tan cattail stalks and murky, coffee-colored water. "Starting with the greening up in the spring, there are seedlings of annual plants sprouting everywhere. Dark green shoots of perennial plants start to poke through the mud. It's like magic, it all happens so fast! By mid to late summer, some plants are eight or ten feet tall, creating a dense wall of deep green," Mary says. The colors change sequentially as different species flower and fade through the season—"the orange of the jewelweed flowers, followed by the spectacular light yellows and greens of the wild rice, and by fall, the deep yellow of the bur marigolds." The whole marsh is awash with color through the seasons.

Mary experienced all this while she pursued treasures hidden in the mud, where a key component of the diversity in a tidal freshwater marsh is found: the "seed bank"—the cache of seeds lying on or just below the surface, waiting to sprout. "Out of a handful of marsh mud, all kinds of interesting plants will grow," she explains. Many of the plants in the tidal freshwater marsh are annuals—they are born, produce seeds, and then die in one year. They take advantage of the patches of open mud that pop up each season. Perennial plants just die back to the roots during the winter (except for woody shrubs and trees, of course).

It was her interest in seeds, and pursuit of the answer to a question, that brought Mary to the tidal marsh in the first place. It all began in mid to late April many years ago, while she was walking in the stream-laced woods near her parents' house in western Massachusetts. There, she encountered by chance many seedlings of jewelweed (*Impatiens capensis*) growing happily along the muddy stream banks. Knowing that most plants have difficulty growing in perpetually wet places because of diminished oxygen levels, she wondered how seeds could have germinated and how the seedlings could survive there. At the end of summer, she returned and collected jewelweed seeds from the spring-loaded seed pods. Living up to another common name, touch-me-not, the ripe green capsules explode at the slightest bump, ejecting their seeds in every direction—an efficient self-dispersal method, and an amusing pastime for wetland explorers, who tend to be easily entertained.

"After collecting the seeds," Mary says, "I stored them carefully on a shelf in the lab. The next June, I was ready to do an experiment, and set out hundreds of seeds in petri dishes. I subjected them to many treatments, to see how they would respond to different light levels, temperatures, pH, et cetera. But only two seeds germinated, out of thousands! Meanwhile, my mom kept some seeds in her refrigerator at home, watering them now and again, and all of hers sprouted in the fridge! This showed me that jewelweed seeds don't tolerate drying, and fresh seeds must be kept cool and moist to germinate. Eventually, I learned what jewelweed seeds needed in order to germinate and went on to learn about the needs of seeds of other species—some need light, some higher temperatures; others, along with jewelweed, germinate in the refrigerator as they would in the spring following low winter temperatures in the ground; still others need to be submerged in water; others need air. The ability of seeds to persist in the seed banks also varies with species; some, like jewelweed, lasted less than one year while others, such as soft rush (*Juncus effusus*) survived burial for decades until brought to the surface."

Jewelweed is common in the tidal marshes along the Delaware River, near Trenton, New Jersey. At the encouragement of one of her colleagues, Mary found herself pulling on hip boots for the first time in her life and walking out into the Hamilton Marsh (now called the Abbott Marshlands).

For more than two decades, she, along with her colleagues and students, surveyed the plants growing in the wetland. They collected and germinated seeds in mud samples, keeping them under carefully controlled conditions in a greenhouse. They discovered that the wetland changed a lot: for example, in some years, there were many bur marigold (*Bidens* spp.) plants, other years only a few or none; and, as a result, the seeds dropping into the sediment changed, too. But overall, the wetland remained the same—the list of plant species didn't change, but the specific locations within the large wetland area did (Parker and Leck 1985).

This result underscores the importance of preserving large, connected networks of wetlands. If the tidal marsh is big enough, every year there will be some place in the wetland with the right conditions for every species, and if it is connected to other wetlands, there will always be sources of new seeds coming into the marsh, contributing to its diversity and beauty for wildlife—and people—to enjoy (Leck 2004; Elsey-Quirk and Leck 2015). "I've brought inner-city kids out to the marsh," Mary says. "Many of them had never been on a field trip before. We got them into hip boots, and once they got their 'marsh legs,' they had a ball. We couldn't pull them out of there— they were just poking around, getting muddy, looking at everything!" Observing the enchantment experienced by these young people, Mary has since spent many years deeply involved in the protection of the Abbott Marshlands and the design of educational materials for its educational center and boardwalk.

Diversity and beauty characterize all these freshwater emergent marshes, whether they are eastern tidal marshes, which experience the ebbing and flood tides, the backwater sloughs along the Mississippi River, or the Iowa ponds and western prairie potholes ringed with cattails and bulrushes. From the shallow edges to the emergent marsh and deepwater aquatic zones, these freshwater marshes are critical places for children to explore, for rails to squawk and toads to trill, for hunters and anglers to pursue bluebills and bluegills, and for biologists to learn how the world works. Mary exclaims, "These places are magical—the experiences they provide are vital. My hope is that young people will visit the marsh often, come to appreciate it on a deep level, and maybe some of them will even pursue a career in biology or, better yet, wetland ecology."

CHAPTER 2

Wet Meadows: Not Too Dry, Not Too Wet

Carex stricta, the firm high place onto which other species cling, avoiding the stress of life below. On the tussock is light, air, a place to call home. It is an ecosystem engineer—organizing the marsh, providing refugia, establishing texture, its form absolutely present in all seasons. A laudable goal to be such a structure in the world outside the marsh.
—SHARON ASHWORTH

Author Catherine writes: Head down, leaning over to fix the water-level recorder, I was hidden behind the tall grasses, sedges, and cattails when I heard the shout over the megaphone. "Michael Reilly, we know you're in the marsh, come out with your hands in the air." I did not know a Michael Reilly, and the only announcements I had ever heard in the 210-acre sedge meadow were the protests of flushed swamp sparrows. Alarmed, I stood up, surprised to see several dozen uniformed officers and police cars lined up along the highway that ran through the wetland. Another four or five men in SWAT jackets patrolled the roof of the shopping center just beyond the road. All startled eyes turned to me; my ridiculous wide-brimmed hat and white T-shirt on my tall frame contrasted sharply against the green sea of dark cattails, pale grass and stripy brown-green sedges. The man with the megaphone, a modern version of Yosemite Sam of cartoon fame—short, stocky, mustached—shouted furiously at me to come straight over to them, *now*. I knew they couldn't get to me in a hurry—moving on the lumpy, wet, and squishy ground, through the tangled mass of plants, was not an easy task, regardless of whether you sported a SWAT jacket. It had taken me a long time to get out there wearing my rubber boots and carrying my equipment, and I could see that I probably wasn't going to be allowed back any time soon, so I leaned over and finished my task.

Strapping on my pack of heavy equipment, I walked the path I had worn over several seasons of work, a meandering path to the edge of the highway; but this wasn't good enough for Yosemite Sam. Blaring through the megaphone, he ordered me to walk straight to them. Stumbling over tussocks, tripping on long cattail blades, blundering into unseen water holes, I attempted a straightaway, knowing full well that I was providing excellent entertainment for the boys in blue. As I walked, I worried—what if this Reilly guy is dangerous? Can I defend myself with a penknife and a tape measure? By the time I arrived at the chain-link fence separating the wetland from the roadbed, I was anxious and the chief officer was apoplectic. "Are you Reilly's girlfriend? What are you doing out there? Did you help him escape?" I assured him that I was merely a graduate student, doing my research, nothing quite so glamorous as a criminal's paramour. He did not seem to believe me and would stand down only after words from his superior officer. After watching me crawl around the end of the fence, the captain assured me that the escaped criminal was unarmed. I told them that my friend Sharon was out in a different section of the marsh, by the river's edge; he directed me to go find her and get her out. As I turned to go, I saw a great blue heron take off from one of the wetter sections of the marsh, near where the railroad crossed; something or someone disturbed that bird . . . "Your man is getting away," I told them, but that particular clue was far too subtle. I easily found Sharon, who, unbeknownst to what appeared to be the city's entire police force, had also stood up from her task, but being shorter than the vegetation had escaped notice. Seeing a distant wall of police on the highway, she had shrugged and proceeded with the tasks at hand along the appointed, well-trodden research route. The day's fieldwork obviously over, we went for ice cream.

Later, I inquired at the local police station and discovered the major transgression that had incited a police dragnet. What horrific crime brought more than thirty officers to the edge of a sedge meadow? Turns out this vicious escaped criminal was wanted for traffic violations.

Walking in the Wet Meadow

If you have ever attempted to traverse an extensive wet meadow like the one described in our story, you might wonder that the criminal never surrendered, begging to be brought to solid, even ground. Without a worn path, you stumble and splash among mounds hidden by dense stands of waist to chest-high grass and long-bladed sedges with sharp edges. But it's worth it. Once upright and stable, you can pause and take in your surroundings. There may be a few clumps of shrubs here and there possibly forming a boundary between your predicament and higher ground. In the distance you may see a northern harrier (*Circus cyaneus*) scouting for white-footed mice (*Peromyscus leucopus*). Stay quiet and listen for a deep clicking followed by what sounds like someone rubbing their fingers over a wet balloon—the northern leopard frog (*Lithobates pipiens*). Returning to the same place mid to late summer rather than spring, rubber boots are no longer necessary, but it is still hard to walk a straight and graceful path. As you blunder, you might flush a gray-chested bird with reddish-brown wings and a matching cap— the swamp sparrow (*Melospiza georgiana*). In the winter, those maddening mounds that tripped you up in July are finally obvious as the vegetation dies back, revealing lumpy ground and an intricate system of tiny, trampled paths through the matted vegetation—meadow vole (*Microtus pennsylvanicus*) runways.

Such hydrologically variable terrains are encountered throughout the United States and go by such names as bluejoint wet meadow, tussock sedge meadow, mixed-graminoid marsh, or simply wet meadow. When dominated by sedges (genus *Carex*), as is common in the Northeast and upper Midwest, these soggy areas are called sedge meadows. On the drier end of the spectrum, a wet meadow in the central United States might be dominated by grasses such as rice cut-grass or prairie cordgrass (*Spartina pectinata*). A wet meadow is defined by a general lack of trees and shrubs, soggy soil, spring flooding, and vegetation consisting mostly of grasses and sedges. They are often found on the edges of freshwater lakes, marshes, and rivers but can form in shallow depressions on the landscape. The lack of standing water for a good portion of the year invites doubt as to the meadow's importance or even its legitimacy as a wetland. To many, it is that no-good swampy part of the pasture, a place easily filled to extend the parking lot, or just that soggy place gone to weeds and shrubs. But hidden under a wet meadow's luxuriously thick carpet of grasses and sedges are nesting places for rails and sparrows, and burrows for such descriptively named creatures as the star-nosed mole (*Condylura cristata*), the meadow jumping mouse (*Zapus hudsonius*), and the short-tailed shrew (*Blarina* sp.). The spring floods turn meadows into nurseries for the amphibian set whose members include northern leopard frogs and blue-spotted salamanders (*Ambystoma laterale*). Not too fond of the twittery, skittery, and slimy? How about clean water and flood-free houses and streets? These meadows filter surface water headed for the river and reduce flooding by slowing and absorbing stormwater.

If you look more closely at a sedge meadow, the plants reveal secrets: they give clues as to the animals that live there, how deep the water gets, and what disturbances have altered the meadow. If you happen to fall to eye level with one of those mounds, there are lessons in hydrology, plant competition, and biodiversity at the tip of your nose.

Not All Wetlands Are for Ducks: Sedges, Soil, and Butterflies

Because meadow wetlands are often perceived as a prelude to "real" wetlands—the ones that contain water and ducks—they are often overlooked. Cast your eyes away from the open water and you will find meadows on the edges of lakes, in association with streams and beaver ponds, or independently established in shallow basins. The sedge meadow will appear as an open grassy area possibly with scattered or clumped shrubs, but no trees. In the absence of mowing or grazing, the paucity of shrubs and trees indicates little disturbance and wet conditions, regardless of whether your feet are damp at the moment. The wet meadow's multiple personalities—from flooded, to saturated, to dry over the course of a growing season, makes it a terribly stressful environment. Red-stemmed red osier dogwood (*Cornus sericea*) and gray-stemmed silky dogwoods (*Cornus amomum*), speckled alders (*Alnus incana*), and narrow-leaved willows (*Salix* spp.) can handle each of these alternate situations; but some common shrubs, like the pom-pom laden buttonbush (*Cephalanthus occidentalis*), tolerate only wetter conditions and so establish themselves closer to permanent water. Other shrubs, like the low-growing, white-flowered meadowsweet (*Spiraea alba*), will handle damp roots but prefer to stay out of flooded areas. Most woody plants found out from under a forest canopy need sunny conditions to ger-minate—a condition not encountered under the thick foliage of the grass-like plants in the meadow. Any hedge you see is likely there because it got lucky: some years ago, seeds dropped by birds or carried by voles landed in a sunny opening (perhaps provided by a cow's heavy footprint, a spot not too wet for too long). After that fortuitous start, shrubs spread by sending out horizontal, ground-level stems and can head out over the meadow if not kept in check by grazing, mowing, burning, or sustained flooding. Meadows overrun by woody interlopers are called shrub-carrs; if you push your way through the dense stems, you can see the remnants of open meadow hum-mocks with their anemic, shaded sprouts of sedge. (Wetlands dominated by shrubs are described further in chapter 5.)

Variably wet conditions also affect the soil in these wetlands. While found atop mineral soils, sedge meadows often have a surface layer—sometimes many inches thick—of well-decomposed organic matter. If you reach down and pick up some of this muck and rub it through your fingers, you won't find big bits of plant matter. *Muck* is not just a name for an expen-sive rain boot, but also the technical term for organic soil in which the dead plant parts are highly decomposed and unrecognizable. Dead plant matter decomposes very slowly when underwater due to the lack of oxygen (see chap. 1, box 2), so continually flooded or saturated wetland soil will have many obvious dead plant parts and is called peat. Dead plant matter in the sedge meadow is alternately inundated by water and exposed to air and so decomposes more rapidly, producing the fine black muck that stains your fingers. Of course, there are murky states of decomposition in between, and you can certainly debate with your friends as to whether the soil beneath your feet, and in which you are potentially stuck, is peaty muck or mucky peat.

After taking in the general openness of the meadow while acknowledg-ing the presence of a few shrubs, the next feature likely to catch your atten-tion is the occasional diffusion of summer flowers into a rough matrix of green—dusty pink joe-pye weed (*Eutrochium maculatum*), most-definitely-gold goldenrod (*Solidago* spp.), lacy white boneset (*Eupatorium perfoliatum*), and the ominous but beautiful purple loosestrife (*Lythrum salicaria*). While the flourishes of color will grab your attention, the dominant pallet of green produced by the humble sedge demands examination.

Not only do sedges look an awful lot like grasses, they look an awful lot like each other. For many specimens in the seemingly vast category of "grass-like plants" a quick check of the shape of the stem distinguishes the sedge—if it's triangular, then it is a sedge ("sedges have edges" as the saying goes—although there are exceptions), a member of the Cyperaceae family, genus *Cyperus* or *Carex*. Attention to this tangled mass of green reveals subtle differences in the color green, plant height, and fineness of the long-bladed leaves. You conclude that there are indeed different species of sedge present, and that may be as far as you get if the plants are not in flower.

Sedges are wind pollinated, their flowers so inconspicuous that it is easy to forget they are actually flowering plants. Like the grasses, they are evolutionary latecomers, their ancestors having given up wind pollination in favor of animal pollination, only to return to reliance on the wind. The male flowers are no more than a small scale (bract) and yellowish stamens that wave in the breeze. The female flowers, just slightly plumper, consist of an ovary enclosed in a sac, called the perigynium, with the arms of the stigma extended out of the sac to catch pollen. No brightly colored petals, no attractive scents—just reproductive efficiency. The resulting seeds not only provide food for small mammals and birds but may possibly be your only hope of identifying the sedge. The fruit of the *Carex* genus is a hard seed called an achene, which is enclosed in the sac-like perigynium and about the size of an unshelled sunflower seed or smaller. The fruits are often bunched together at the end of the stem. Some have protrusions, resulting in a stiff-bristle-bottle-cleaner look; some bend the end of the stem and are reminiscent of pendulous catkins; and still others look like three-dimensional origami stars.

Sedges are common in all types of wetlands, not only sedge meadows. What is it about their structure or ecology that makes them so well adapted for wet situations? Like many wetland-adapted plants, *Carex* have long, underground root structures called rhizomes, which allow them to move into wet situations quickly. Abundant aerenchyma—air-filled "pipes" that pump air from stems to roots—occupy the plant's roots, rhizomes, and stems (see chap. 1, box 2). The sacs around the seeds, the perigynia, trap air, allowing the seed to float and disperse quickly during high water. Additionally, the seeds stick to the feet and survive in the guts of ducks, muskrats, and even elk, allowing *Carex* to disperse far and wide—even between continents (Waterway, Hoshino, and Masaki 2009). Perhaps botanist and author Linda Curtis of Lake Villa, Illinois, summarizes it best: "Cyperaceae are the third largest family globally, and genus Carex owes its success to its variability, which also makes identification so problematic" (Curtis 2016). Linda describes the source of *Carex*'s variability as the "oops, slips, reverses and duplications" of chromosomes—a rather more poetic way of describing duplication and inheritance of chromosome fragments resulting from holocentric (nonlocalized) centromeres (Hipp, Rothrock, and Roalson 2009). Linda Curtis has written numerous books on sedges, identifying the *Carex* genus in special places in Florida, Illinois, and Wisconsin; but the title of one of her essays might catch your attention: "What Good Are Sedges?" (see Curtis 2016). It's a fair question, and one often posed by those who find places where sedges grow uninteresting at best, or an unnecessary barrier to development at worst.

There are a host of ecological lessons that can be illustrated by examining some sedges up close. Spend some time flipping over the leaves of sedges and look for pale-green spheres about the size of a single BB or tapioca pearl (some of these delicate spheres may sport two brown concentric circles). These are the ova (eggs) of the Dion skipper (*Euphyes dion*), a rusty-brown, fuzzy-bodied butterfly with small yellow splotches on its wings. Maybe these tiny peas have irregular, faded brown spots, in which case they are the ova of the broad-winged skipper (*Poanes viator viator*), another butterfly that you might also describe as above, except with larger yellow-orange splotches. The caterpillars of these two native butterfly species depend on sedges, particularly lake sedge (*Carex lacustris*), for food (Shapiro 1970; Shuey 1993). If the tiny spheres you see on the underside of sedge leaves are white, they may be the eggs of the mulberry wing (*Poanes massasoit*), a butterfly dependent on the tussock sedge (*Carex stricta*) (Lotts and Naberhaus 2014). When these minuscule eggs hatch, the fingernail-size caterpillars will feast only on the leaves of these particular plant species, often using a bit of silk to create a little gossamer tent between the stem and leaf.

Much like drier meadows and old fields, wet meadows can be critical habitat for native pollinator populations. Not only do the colorful prima donna flowers of the wet meadow described earlier provide nectar for butterflies and moths, certain pollinators depend on the backdrop of wetland sedges as host plants for the larval or caterpillar stage. In fact, all the northeastern US *Euphyes* butterflies—black dash (*Euphyes conspicua*) and the two-spotted skipper (*Euphyes bimacula*), in addition to the Dion and the broad-winged—depend on sedges as larval hosts (Shuey 1993; Kart et al. 2005). As far as we know, the existence of these butterflies is not crucial to the pollination of food crops, they are not the key item in the diet of marsh carnivores, and they do not play a role in the control of unwanted plant species; they are valuable simply because they are part of the complex food web of sedge wetlands and are disappearing along with the only habitat they are known to depend on.

Moving Up and Out: Tussock Sedge as Ecosystem Engineer

Now let's focus for a minute on one of these sedges—the one that is responsible for all those mounds Catherine tripped over attempting to extricate herself from the meadow under police orders, and that you probably tripped over while looking for butterfly eggs. The tussock sedge is not as it appears—a plant growing on a mound of muck—but rather the mound is the plant, its roots, stems, and leaves along with some collected soil and dead plant material. This sedge species is the engineer of the meadow, providing refuge from floodwaters and organizing the meadow environment by moisture. The mounds sedges create give the meadow its topographic texture, their form present in all seasons—appearing like goosebumps on the winter landscape, like frightened cartoon characters with bright green spikes growing from the top of the mound in spring, and like flowing, grassy bouffant hair pieces in the summertime. The seed heads of tussock sedge are long and narrow, the male flowers on separate spikes above the female flowers. Each flower has a rich red-brown scale, giving the female spikes a contrasting pattern with the light-green perigynia. The tussocks can range from just above ground level to just over three feet tall; the taller and more voluminous the tussock, the deeper and more prolonged the flooding in the meadow (Lawrence and Zedler 2011). This vertical variety illustrates the tussock sedge's aptitude for keeping some roots and stems above water. The roots, called adventitious roots, grow aboveground along the length of the stem in addition to below-ground at the base of the stem. Roots need oxygen to function, which is why it is more proper to think of wetland plants as tolerating rather than prefer-ring oxygen-deprived, wet conditions. Adventitious roots allow the inun-dated plant to avoid low-oxygen conditions by simply growing up and out of the water, increasing the size and height of the tussock, and exposing tissue to an oxygen-rich environment. In many wetland plants, including tussock sedge, these roots and stems contain aerenchyma, a spongy tissue filled with air spaces that allows oxygen to diffuse from aerial portions to plant parts underwater, acting as a kind of snorkel for the plant (see chap. 1, box 2).

Looking more closely at the mounds and comparing them to the sur-rounding space between them, you are apt to find other vegetative residents tucked in among the leaves like high-rise apartment dwellers. The mounds and the intermound spaces, just centimeters from each other, constitute dif-ferent habitats. Compared to equivalently sized flat places, tussock mounds have greater microhabitat diversity—small differences in light and moisture creating different living conditions or niches. Tussock sedge is the high, firm place onto which other species cling to avoid the stress of oxygen-deprived life in the water below. The tussock is where there is light, air, and a dry place; because of that, there are a greater variety of plants making their homes on the tussocks than between the tussocks. And the taller the tussock, the more spaces available, and therefore the greater variety of species you'll encounter (Peach and Zedler 2006). The top of the tussock is driest and has the best access to light and oxygen (Bledsoe and Shear 2000), and during flooding it's the best spot around—the penthouse, if you will. As floodwaters recede, more spaces open up along the sides of the tussocks, balconies on the side of the high-rise. It's shady under last year's sedge leaves, but there is less competition for these spots, and some plants get just enough sun to thrive. Research in Wisconsin identified at least twenty-nine species that reside on tussocks—most preferring to establish themselves on the top, some accept-ing various locations along the sides, and a few able to tolerate the bottom floors (Peach and Zedler 2006). Certainly not all tussocks are loaded with other occupants; however, the diversity of habitat provided by a collection of mounds is greater than might appear at first glance. While there is space to live in between the tussocks, the basement is always flooded, the tempera-ture is colder (Peach and Zedler 2006), access to light is limited by an accumulation of past years' dead and dying tussock residents, there is no view, and the four-footed traffic is a nightmare (Crain and Bertness 2005).

In drier sedge meadows, the physical and botanical contrast between tussock and non-tussock is not so stark. Shallower, shorter periods of flooding lessen the need for dramatic building operations to keep above water. In these meadows, tussock sedge is not so domineering, making way for other sedges such as the tall, fine-leaved water sedge (*Carex aquatilis*) and the stout, wide-leaved lake sedge. Some species may be indicators of soil acidity. More-alkaline (less-acid) areas may include calcium-loving species like the red-footed spike rush (*Eleocharis erythropoda*), bristle-stalk sedge (*Carex leptalea*), rigid sedge (*Carex tetanica*), and swollen-beaked sedge (*Carex utriculata*) (New England Wildflower Society, n.d.). You will also notice more grasses (round-stemmed "grass-like" plants that are actual grasses) and forbs (pretty flowers) in the matrix of plants. A commonly associated grass is *Calamagrostis canadensis*, or bluejoint grass. As its name implies, the three-foot-tall, light-green grass has darker, bluish-purple joints in addition to a cream-colored inflorescence. Rice cut-grass (*Leersia oryzoides*) may be another notable addition, and one that is hard to miss if it is present. This fine-textured grass with its fuzzy joints may be overlooked at first in a sea of green, but wander into a patch with shorts on and you'll end up with a multitude of scratches; run your hand along the leaf edge and you can end up with paper-cut lacerations. As unpleasant as rice cut-grass can be, its seeds feed ducks and other wetland birds such as the swamp sparrow.

The fascinating, frustrating, wondrous thing about wet meadows is their subtle and informatory patchiness. It may be difficult to walk a straight line, but you can end up stumbling through plants that reveal the wetland's hydrology, soil diagnostics, animal life, and microtopography. The plants also create hidden environments that not only harbor small creatures, but also create a giant water purification factory underground.

More Than Meets the Eye: Wet Meadows and Water Quality

The sedge meadow in the opening story is an urban wetland, surrounded on the upland side by the pollutant-producing pavement of parking lots, roads, and shopping malls. On the downstream side, beyond the sedges and grasses hiding the criminal, is the Yahara River. The Yahara is an urban river connecting several water bodies in and around Madison, Wisconsin. From that urban area washes great quantities of street and yard detritus—oil, gas, garbage, pesticides, and sediment. Should that bilge, or the stormwater from the uplands, pass through the wetland, the wetland will act as a filter, and the water entering the river will be cleaner for it.

Filtration happens aboveground as well as belowground, where a dense network of fine plant roots, beneficial bacteria and fungi, and porous organic soil captures pollutants. As water flows in from the uplands, plant stems, plant roots, and the bumps of soil and debris slow the water down, causing sediment to drop out of the water column, settle onto the surface of the wetland, and mingle with the wetland soil below. At the same time, heavy metals in the water—toxic elements such as copper, aluminum, and lead—bind to the soil particles (Chen 2011). Complex chemicals such as pesticides, grease, and gasoline are decomposed into less harmful constituents by wetland microbiota.

Two common pollutants, nitrogen and phosphorous, are washed in not only from urban landscapes but agricultural landscapes as well. Both nitrogen and phosphorus are naturally scarce (limiting) nutrients in aquatic systems, so augmenting them will incite a riot of plant growth in any aquatic system (although phosphorous tends to be more limiting than nitrogen). Too much nitrogen and phosphorus can cause explosive algae growth, and when all that algae dies, decomposers deplete available oxygen. When oxygen levels plummet, so do populations of frogs, clams, aquatic insects, fish—the whole food web can go down. Nitrogen and phosphorus are added to lawns, gardens, and farms as fertilizer, often in higher amounts than can be absorbed at those places; the excess washes downstream with the rain. These pollutants are also washed downstream from excrement of any kind—dog, cow, and waterfowl are common sources. Human waste can add to the pollutant burden through wastewater discharge from a sewage treatment plant or septic system.

Fortunately, wetlands are well positioned to deal with these overabundant nutrients. Standing between open waters and agricultural land or suburban lawns, shallow marshes and wet meadows suck up and store phosphorous and nitrogen carried on sediments and dissolved in water after a rainfall on the fields (J. Zedler 2003; Hoffmann et al. 2009). The alchemy that wet-lands use to convert harmful, waterborne nitrogen into harmless nitrogen gas is nothing short of miraculous.

The process starts when one form of nitrogen (called nitrate) flows into the wet meadow. Nitrate dissolves easily in water and is happily slurped up by the vegetation. Much of that nitrogen is released back into the water once the plants die, but often at a slower rate, and some of the nitrogen is "locked up" in the undecayed plant parts that make up the organic soil (remember, in wet situations, there isn't much decomposition, so pieces of leaves, roots, stems, etc., just drop to the ground and become part of the soil, taking the nitrogen with them). What isn't taken up by plants and converted to new plant growth is converted into a gaseous form, ultimately molecular nitro-gen (N_2), by denitrifying bacteria in the soil and then released into the air. Since almost 80% of the air we breathe every day is made of nitrogen gas, this process, called denitrification, is completely natural and harmless. All wetlands perform denitrification, but the wetlands that do it best have both flooded and unflooded places, dense vegetation, and organic soils; a sedge meadow.

Getting rid of the phosphorus is much trickier. Plants take some of it in, but most of that is released when the plants die back for winter. Fortunately, phosphorus will form a chemical complex with iron or other elements, and when this happens the phosphorus is rendered insoluble and remains in the soil, harmless. However, this process requires oxygen. If the water is deep and stagnant—for example, in wetlands that are wet all the time—there won't be enough oxygen and the phosphorus will not remain in the sedi-ments. Thus, the drier portions of wetlands, or wetlands with sufficient water movement, will have enough oxygen to retain phosphorous.

Where does the nitrogen and phosphorus end up should there be no wetlands to capture it? If you are in the middle of the United States, these nutrients make their way down the main drainage path, the Mississippi, and out into the Gulf of Mexico where excess nitrogen spurs algae growth, lead-ing to eutrophication and hypoxia (lack of oxygen). William Mitsch and col-leagues (2005) estimate that it would take a wetland creation and restora-tion effort resulting in twenty-two thousand square kilometers (8,494 sq. miles) of wetland (about the size of New Jersey) to reduce nitrate-nitrogen runoff into the Gulf of Mexico by 40%. Wetland loss in the Mississippi River basin, where three-quarters of all original wetlands have been destroyed, is directly linked to the infamous "dead zone" in the Gulf. On the East Coast, Chesapeake Bay is suffering in much the same way — extensive dead zones, killing the famed blue crabs and oysters that feed happy tourists and support so many fishing families and local economies. The decline in water quality in these important bodies of water is directly linked to the loss of wetlands such as sedge meadows.

Unfortunately, the sedge meadow may ultimately sacrifice itself in service to water quality. There is only so much chemical insult a wetland can take, and excess nitrogen and phosphorus may ultimately lead to the invasion of the meadows by unwanted intruders.

The Ballast Waif and the Gardener's Garters: The Tale of Two Invasive Species

The presence of two particular plants in a meadow reveals a great deal of information — none of it good — about the condition of the wetland. Purple loosestrife and reed canary grass are poster plants for invasive species biology. Each has an arsenal of strategies to take advantage of the slightest opening, establish themselves as real estate tycoons, and then bully, beat, and baffle attempts at eradication. In the case of the purple beauty, however, there is some recent doubt about its reputation as a wetland killer, while at the same time there is not enough attention paid to the more insidious invader, reed canary grass.

Purple loosestrife (*Lythrum salicaria*), an exotic lovely from Europe, likely came to early nineteenth-century New England shores in ship ballast, joining a growing list of "ballast waifs" (Thompson, Stuckey, and Thompson 1987) immigrating to the New World. The seeds also came with four-legged and two-legged immigrants — hitching rides on the wool of sheep imported to New England's woolen mills (Stuckey 1980, cited in Thompson, Stuckey, and Thompson 1987) and brought by people who thought they carried with them a treatment for dysentery and an antiseptic for wounds and sores. The plant's beautiful spikes of purple-pink flowers also made it a sought-after horticultural specimen for gardens. But, like many nonnative species, it escaped its domesticated settings. Purple loosestrife's path to the wild, and to the rest of the country, was predominantly the canals and waterways that pushed inland during the 1880s. Further spread has been aided and abetted by its use in gardens and wildflower mixes.

Purple loosestrife has a reputation for crowding out native species and taking over (Thompson, Stuckey, and Thompson 1987; Mal et al. 1992; Blossey, Skinner, and Taylor 2001; Schooler, McEvoy, and Coombs 2006)—its spread certainly looks dramatic with those attention commanding blossoms—but it wasn't until a 1987 US Fish and Wildlife publication (see Thompson, Stuckey, and Thompson 1987) sounded the alarm that scientists and the public really took notice (see Lavoie 2010). The alarm launched a public campaign to control purple loosestrife, and the popular press rallied, referring to the lovely loosestrife by unappealing monikers such as *invader* and *menace*, and in a few cases *thug, monster, nightmare*, and *barbarian* (Lavoie 2010). It is listed as a noxious weed in thirty states (US Department of Agriculture, n.d.). Purple loosestrife certainly has the armory of a potential catastrophic invader, producing nearly twenty-two thousand seeds the size of sand grains for each flowering spike (Lindgren and Walker 2012). Once released, the seeds fall to the mud below and can be carried away on boot, paw, hoof, or tire to the next patch of moist ground. Purple loosestrife is not picky about soil type, so any moist ground will do, and once established will be terribly difficult to get rid of. By all means, do try to pull the plant out if you encounter it, but make sure you get all of it—any remaining stems or roots will resprout. That perennial rootstock will also recover after the aboveground parts are doused with herbicides. Use a herbicide that will affect the roots and you are likely to do damage to the rest of the wetland inhabitants—not to mention you will have to hit the offending plants year after year after year to see any affect. Burning, mowing, flooding—all are for naught; the illustrious infestation will not disappear unless it is eaten. While adding purple loosestrife to your bowl of salad greens won't work, the plant is definitely on the menu of certain leaf-eating beetles.

Herein lies a tricky ecological dilemma: Do you introduce yet another exotic species to control the one that got away? In addition to thousands of tiny seeds and stubborn rootstocks, another reason purple loosestrife found America such an accommodating place is the fact that nothing here would eat it. Back home in Europe are voracious leaf-eating beetles, root-mining weevils, seed-eating weevils, and flower-eating weevils among a host of insects dining on purple loosestrife (Blossey 2002). Bringing foreign insects to the United States can introduce biological control agents to address the spread of unwanted interlopers — or just create another ecological nightmare. Years of carefully monitoring limited introductions are warranted in such cases. After determining that they would not be a serious threat to native plants or to crop plants, two leaf-eating beetles and the root-mining weevil were chosen and introduced to their new homeland in 1992. Happy to find a native dish in plentiful supply, the insects munched away and appear to be successfully holding purple loosestrife populations in check (Blossey 2002; but see Grevstad 2006; Hinz 2014).

The cost of evaluating an introduction of an exotic species and the subsequent follow-up is considerable, and so the question of the severity of the initial threat is relevant. Despite purple loosestrife's biology, its reputation as a home wrecker, and the efforts that have gone into controlling the plant, there is some recent evidence to suggest that loosestrife might not be the scourge once thought — or, at least, does not present a death sentence to wetlands. A few studies submit that invading purple loosestrife does not reduce the diversity of other flora species (Treberg and Husband 1999; Farnsworth and Ellis 2001; Morrison 2002; Hager and Vinebrooke 2004), at least before loosestrife reaches some critical biomass (Farnsworth and Ellis 2001). It may be that, in the 150 years that purple loosestrife has been present in the northeastern states, it has come to be less of a threat in this area of the country as large infestations become less common (Lavoie 2010). While the debate rages on, early control and management of purple loosestrife where possible is certainly desirable, as attempts at eradicating large infestations are potentially expensive and problematic. One of the most distressing findings of some of the research has been that as the density of purple loosestrife declines, the presence of another wetland invader increases (Morrison 2002; Schooler, McEvoy, and Coombs 2006) — one whose effects are indisputably hostile.

Gardener's garters, more commonly known as reed canary grass (*Phalaris arundinacea*), is routinely offered in seed catalogs. A popular forage planting, it is favored for erosion control, useful for the treatment of wastewater, and touted as a biofuel.

Reed canary grass is a tall-growing, perennial grass that is widely distributed across Minnesota and other northern states. Particularly well adapted to wet soils, it is also productive on upland sites. Reed canary grass spreads by underground stems (rhizomes) and forms a solid sod. It can be harvested as pasture, silage, or hay, whether sown in pure stands or in mixture with legumes. (An entry from the Hancock Seed Company's 2018 online catalog, https://hancockseed.com/reed-canary-grass-seed-25-lb-bag -1010.html)

If you see reed canary grass in a wet meadow, something has gone terribly wrong. Typically found in dense, single-species patches, this tall grass sports leaves that stick out in all directions—leaves much shorter than the plant is tall—and straw yellow seed heads that haughtily wave above it all. Imagine what your unmowed lawn would look like waist- to shoulder-high and with fewer weeds. Like kudzu, the vine that ate the South, reed canary is the grass that ate the wetland. Unlike its partner in crime, the deceptively beautiful purple loosestrife, reed canary grass leaves no doubt that it has taken over the wetland to the detriment of any and all resident species, and there is nothing attractive about it. Isabel Rojas, who as a graduate student at the University of Wisconsin–Madison did research in sedge meadows, remarked on how lovely the meadows were, especially when plants such as milkweed, tucked between the sedges, were blooming. "But right next [to the sedges], in the *Phalaris* patch you won't see anything—it's so much biomass. It's so impressive how dense it is," she says. "There maybe was one other plant [in the reed canary grass stand], one tiny jewelweed seedling."

The exact origin of reed canary grass is unknown, as it has long been cultivated in northern temperate climates around the globe. The Swedish began cultivating it for forage as far back as 1749, and it was cultivated in Connecticut and New Hampshire in 1834 and 1835 (Galatowitsch et al. 1999). There is a relatively well-behaved native form of the grass in the Pacific Northwest, but the problem appears to stem from a number of introduced cultivated varieties originating in Europe: aggressive cultivars introduced for forage and erosion control that spread from agricultural ditches into meadows, swallowing them whole. (See Jakubowski, Casler, and Jackson 2012 for a detailed analysis of reed canary grass origins.)

As with many invasions, there is often an inadvertent invitation. In the case of reed canary grass, a perfectly nice meadow accumulates sediment from stormwater runoff, or receives an influx of nutrients from fertilizers applied to neighboring farm fields and lawns. The grass seed, waterborne from an overflowing ditch, settles in the meadow; an opening in the canopy of sedges allows light to shine on the moist soil; and the infestation begins. Once established, you can actually watch an advancing frontline of reed canary grass invade a meadow, no seeds or sun necessary. Under your feet is the battle for new ground, the grass sending out rhizomes and tillers into virgin territory. These explorers receive all necessary supplies from the mother plant and so can survive low levels of food, water, and light in the new terrain—in fact, the clones that sprout from these invading scouts will tolerate varying water levels and nutrient quantities that the natives will not (Lavergne and Molofsky 2004; Maurer and Zedler 2002).

The new sprouts sent up from the creeping rhizomes will green faster than the native sedges in the spring, getting a head start in the critical competition for space. The new recruits then elongate rapidly to capture the sunlight and shade their neighbors. Should the conditions of battle change, the field beset by drought or flood, the reed canary grass can quickly adapt its physical structure, allocating more resources to the stems and leaves aboveground or to the roots and rhizomes belowground as needed (Lavergne and Molofsky 2004). Should the resident plants obtain enough sunlight to make a stand, reed canary grass, having reached its full height and mass at the end of the summer, will simply fall over on its neighbors (Healy and Zedler 2010), definitively ending such profligate use of the limited resource. Leaving no room for doubt as to its superiority, the marauding grass remains green long after other plants have senesced, storing supplies for winter and another overwhelming assault come spring (Lavergne and Molofsky 2004).

These invasions are of the most insidious sort—the plant moves in and changes the environment, promoting the increase of its own kind. A dense stand of reed canary grass will capture more sediment from water running off the landscape than will the sedges, promoting favorable conditions for its continued spread (Bernard and Lauve 1995). And there is no doubt that the transformed landscape is devoid of plant diversity—species decline is linked directly to the increased presence of reed canary grass (Werner and Zedler 2002). The pinks, whites, and golds of meadow flowers disappear, to be replaced by one shade of green. The variety of sedge shapes and sizes is gone, replaced by a single indistinct flag. The long, graceful ribbons of sedge leaves, the tall tufted heads of rushes, the feathery marsh ferns — all gone — replaced by a uniform stand of grass.

Aficionados of the sedge meadow are many, however, so this invasion is not taken lightly. In Wisconsin, the husband-wife team of Dan Collins and Nancy Aten describe themselves as reed canary grass predators. While you may cock your head upon first hearing the term *predator* used in this context, bear in mind that, unlike purple loosestrife, reed canary grass has no six-legged predator on this continent; so two-legged predators will have to suffice. The term's appropriateness becomes clear as Dan describes their methods: stalk, extract, decapitate, and, finally, apply the "glove of death."

Stalking involves paying deep attention to the plant and the site, determining when growth begins, when pollen is released, when seeds begin to shatter. "Those points in phenology are inflection points for a process to help rid the wetland [of reed canary grass]," Dan explains. Then there is the counting: how many fruiting stems, how many seeds per stem. Yes, Dan knows the average seeds per stem at his current project site — 60 to 160 seeds per stem. Knowing these details "tells you what sort of expectation you have for getting rid of it," Dan continues (although the phrase "getting rid of it" is rather loose terminology). The hunting season begins with extraction, pulling out as many of the plants as possible (this is more easily done in inundated sites), followed by decapitation and bagging before the seeds are released. With a pair of scissors and three hours, Dan can remove a quarter million seeds! Tackling an infestation for the first time may be daunting, netting about three thousand stems over a couple of hundred square yards, but once an area is treated this intensively, Dan says, you can work more than a half acre in maintenance mode. Of course, the battle is not over even after a decapitation. And so Dan brings out the "glove of death," an herbicide-soaked cotton glove that he wears over a protective glove and uses to grasp the plant, coating it with the killing chemical.

Such attention can mean up to fifteen hours per acre at the start of a wetland restoration effort, but will taper off to maybe an hour per acre if vigilance is maintained every year. With reed canary reduced, Dan and Nancy find that the wetland begins to work with them — the natives surge and take their rightful place. Such dogged persistence has its rewards. "It gets very zen-like after a while," Dan says. "If you enjoy being out-of-doors and you

like being in wetlands, then this is a great way to be out of doors in a wetland for a lot of hours." Quiet hours spent killing reed canary has taught Dan
to look out for suspicious hubcaps—not road trash, as first suspected, but
snapping turtles. He's also intimately familiar with the rhythm of life in a
wetland, the cycles of emergence, breeding, nesting, spawning, and hatching. And almost exclusive to the patient wetland wader are opportunities
to spy shy fauna like the elusive American bittern. Most rewarding is that
removing reed canary grass can produce dramatic results, and once again
the meadow will sport splashes of color and attendant butterflies in nests of
graceful sedge.

CHAPTER 3

Pond-Meadow-Forest, Repeat: The Beaver's Tale

Beavers are the great comeback story, a species that outlasted the Ice Age, major droughts, the fur trade, urbanization and near extinction. It is one of the few species that can go head to head with humans and win. And so the battle for world domination begins.

–GLYNNIS HOOD, *The Beaver Manifesto*

In 1985, when he first moved to Westmoreland, New Hampshire, John R. Harris spent many afternoons walking in the woods and fields behind his house following deer and fox tracks, noting where birds had built their nests and watching for garter snakes sunning in the late-summer heat. An undergraduate chemistry major turned English professor, John nursed a keen interest in the intersection of cultural and natural history that visibly graces so many New England properties. His explorations brought him to cellar holes and abandoned wells along woodland roads, and ultimately to an extensive wetland several miles east of town. There he watched beavers as they rose to the surface at twilight to repair their dams, harvest saplings, and interact with one another along the shore. One evening a neighbor, Linwood Burt, stopped to talk. "He said that he knew the place well, for it figured prominently in Westmoreland's earliest history," John recalls. "According to Linwood, the town's first arrivals, who settled along the eastern bank of the Connecticut River in 1741, brought with them a single horse. Somehow that summer, this valuable animal escaped, and was presumed lost by the dozen frontier settlers. The following spring, one of the settlers spotted the runaway horse browsing in a lush hay meadow at the base of Seventy-Acre Hill. He approached, coaxed the animal under his lead, and returned to the crude cabins they had built along the river. Before he left, however, in acknowledgment of his great good fortune, this man named the place where the horse was recaptured the Lord's Meadow."

When he returned to the Lord's Meadow wetland the next day, John thought about Linwood's story, marveling that the beaver pond area was once dry enough for a horse to graze there. He then noticed features he had earlier ignored, such as a stone wall that vanished into the pond, indicating that the area used to be a meadow, with grazing animals kept in by the barrier of stones. "In place of the waist-deep water I was seeing now, I tried to imagine a twenty-acre field of sedge and grass surrounded by forest. I wondered how long ago the beavers had moved in and reclaimed this low-lying area that had originally been their domain." His inquiries led him to Mary Fredette, a sprightly eighty-year-old retired postmistress who lived in a small red Cape Cod abutting the pond. "She vividly remembered the year, thirty years prior, when the beavers had taken up residence at the pond, as well as two years earlier, in 1956, when a couple of local farm boys had been forced to abandon their wagon filled with hay after it became mired above the axels in September," John says. The wetland was dry enough to be used as a hay meadow because the beavers had left; but when they returned, the area became too waterlogged for farm equipment. Today, the wetland continues to support three active lodges and at least nine beavers.

Learning the name of the wetland and understanding its history changed John's relationship with the Lord's Meadow. "The area so clearly shows a reciprocity between humans and nonhuman nature," John explains. "William Cronon, in his book *Changes in the Land*, discusses this two-way relationship—the landscape we see today is the result of the interplay between cultural and natural forces." Beavers change the land to suit themselves, and so do people, leaving traces of their work long after they are gone. The effects of Native Americans, early settlers, and wildlife are all visible in the environment—although in New England it is all overgrown by forest, so you have to know how to look for it.

Here and Gone: The Disappearance of the Beaver from North America

The Lord's Meadow serves as a lovely illustration of the cycle of pond-meadow-forest that comes in the wake of beaver (*Castor canadensis*) colonization, construction, and abandonment. Surveying what seems to be a wet meadow, such as the one where the lost horse was found, there are often many signs that beavers were once present. The old beaver dams may persist as low, serpentine mud walls, covered in short sedges and delicate bugle-weeds (*Lycopus* spp.), winding their way through shrubs. Upstream from the walls, large earthen hillocks protrude from the level landscape, most sprouting marsh ferns (*Thelypteris palustris*), buckthorn (*Frangula alnus*), and speckled alder shrubs (*Alnus incana*)—these are the lodges, many years abandoned. Pointed stumps hide among the grasses and sedges, evidence of woody plants gnawed down by beavers. Everywhere, the driftwood-gray dead trees—killed by high water many, many years ago when the dams were built—stand in marked contrast to the yellow-and-green-striped texture of tall grasses and sedges in the background.

This cycle of pond-meadow-forest created by beavers demonstrates well the complex and dynamic nature of wetlands. Subject to vagaries of weather, wildlife, and water, wetlands are constantly changing, as one species of plant or animal takes advantage of new resources available, creating a new wetland type. The beaver is one of the most common and most powerful agents of change in a wetland.

Prior to 1700, every pond, wet meadow, or streamside flat in North America would have boasted at least one beaver family. At that time, when the first white settlers surveyed the land, recording witness trees and stream-crossed meadows, the European desire for top hats made of lustrously warm, felted beaver fur created a lucrative market for trappers. The musky castor oil secreted by these large rodents was also collected and sold, for use in the perfume industry. By submersing metal spring-jaw traps in ponds and waterways, fur takers left the North American landscape virtually devoid of beavers by 1900.

Even after the beavers were gone, their legacy remained. Recognizing the value of the wet meadows created by the absence of beavers, farmers in the hill country of New Hampshire, Vermont, Massachusetts, and New York, as well as in the fertile farmlands of the Midwest, would cut hay in these newly beaver-free wet meadows, keeping trees and shrubs from growing back. These hayfields benefited from the overflowing streams, which brought nutrients washed downstream from the hillsides, feeding lush green forage for sheep and cows aplenty. Country people, like the settlers along the Connecticut River, knew the importance of these meadows and tried to incorporate at least one streamside meadow into their plot (Donahue 2004). Often, cart roads were built across the tops of former beaver dams, and stone culverts were constructed underneath to keep water flowing through and the meadow just moist enough. Beavers, being efficient, had often chosen the narrowest point of the stream for their dams; farmers, being equally practical, followed suit.

Beavers are often lauded as nature's "ecological engineer," one of the very few species other than humans that can drastically alter the landscape. Employing extensive knowledge of construction, silviculture, and plumbing, beavers act as a keystone species, creating a diverse set of physical conditions that support a wide variety of animal and plant species. Without maintenance from beavers, the dams they build eventually collapse, the area behind the dams dries out, and shrubs and other woody plants move back in. Eventually, the forest regrows and the meadows and ponds disappear. With beavers, the river valley sported ponds, deep marshes, and muddy meadows as well as free-flowing stream segments; without beavers, the trees crowded in, the river corridor narrowed, and the landscape lost some of its diversity (Naiman, Johnston, and Kelley 1988; Burchsted et al. 2010).

The near extinction of the beaver in the earlier part of the twentieth century left behind a drier landscape, with fewer fish, amphibians, waterfowl, and aquatic insects. Realizing that without the ponds and pools created by beavers, there would be fewer places to fish and fewer fish to catch, game wardens, trappers, and ecologists conspired to bring the beavers back. The reintroduction of beavers, from 1930 to 1950, is a story repeated across the country in many different forms. One of the more adventurous tales is re-counted in Jennifer Lovett's children's book, *Beavers Away!*, which takes place in Idaho. In most places, bringing the beavers back was just a matter of livetrapping several young males and females of the species and driving them to the nearest free-flowing stream. In the steep hills of Idaho in 1947, however, it was not so easy — it took two days of riding hot, dusty trails on horses or mules easily spooked by the odorous, upset beavers strapped to their backs; then, many miles by truck; and at length again on horseback for the final leg of the trip.

So many beavers overheated and died in this process that a game warden named Elmo Heter felt compelled to find a different way. Using his skills as a bush pilot, he assembled a parachute system to safely deliver young bea-vers to their new homes. One unlucky beaver, aptly named Geronimo, was drafted for repeated tests of this airplane-ejection-soft-landing technique: he'd be dropped out of the plane in a cage and float down to the ground on the parachute; then the cage would pop open. He'd scurry for freedom — only to be recaptured by waiting conservation officers and put through the same routine again. After several unsuccessful postdrop attempts to escape, Geronimo would return, with no urging, to the livetrap, resigned to his fate as a furry paratrooper for science. Once the airborne introduction tech-nique was perfected, Geronimo and three young female beavers were set free; seventy-nine beavers were released throughout the mountains this way, waddling off to explore their new territory and establish themselves in the uninhabited wilderness (Heter 1950).

These newly landed beavers were faced with abundant undammed streams, but many of the streamside habitats had grown in with hemlocks, ash, red maple, and other less desirable tree species. Fortunately, evolution left the beaver with a set of features and instincts keenly honed to turning an inhospitable site into an aquatic paradise. Among them: sharp front teeth that grow incessantly unless worn down by the action of gnawing wood; fur-lined lips that close behind the teeth for underwater chewing; a long flat tail used not only for swimming, but also for temperature regulation, stability on land, fat storage, and a noisy slap-on-the-water warning system; a special-ized digestive tract enabling a high-fiber diet; and an irrepressible instinct to build dams (Müller-Schwarze and Sun 2003).

Plumber, Lumberjack, Builder: All in a Day's Work

To a beaver, the sound of running water is a call to action: Turn off the water, stop the flow. Build. Build quickly. Build a dam. One of the most important reasons for the pond created by the dam is to provide cold storage under the ice surface for the beavers' winter diet of sticks and twigs. Beavers can't walk easily on land, ice, or snow; they prefer to swim to their food all year round. The high water behind the dam floods more land, thereby providing watery pathways to a larger number of tasty trees. Using their self-sharpening and quick-growing front teeth, the beavers will cut the most nutritious trees first to get to the edible branches. The cambium, or inner layer of bark, constitutes a tasty staple of their diet, but they also eat young twigs in winter; grasses, ferns, and wildflowers in spring and fall; and aquatic plants in summer (Rosell et al. 2005). Since beavers can't climb, they have to cut the trees down to get to the young twigs. They cut and store a good portion of the smaller branches in underwater caches, accessible under the ice in winter. Other parts of the felled tree are used to build the dam and the lodge.

The pond also creates perfect growing conditions for their summer food supply of pondweeds and water lilies. The lodge, surrounded by the deep water of the pond and entered only from underwater, is protected from most predators, and the pond itself makes it easier for the beavers to escape pursuers seeking to do them harm. In addition, the dam slows down the flow of water, collecting an impressive amount of woody debris and sediment, and the pond itself constitutes prime habitat for an astounding number of water-loving creatures.

To build their dams, beavers push cut twigs and branches into the bottom of sediments, anchoring them with stones. Then they dig up mud and carry it with their front paws, placing the material between the branches to cement the wall of the dam. Grass and more branches are piled or interwoven into the structure (Müller-Schwarze and Sun 2003). Dams may be built with the inside curve facing upstream or downstream, or they may be straight across. Many dams are built in small sections, connecting protruding rocks or tree hummocks. In small streams, the result is a low wall of sticks, mud, and rocks, often topped with seedlings of grasses and sedges. Where the flow is greater, the dam is more impressive — a three- or four-foot wall of sturdy sticks, mortared with plenty of dark mud.

Every night, the beavers inspect the dam for leaks, repeating their mud-masonry endeavors as needed. The stream often continues to run over the top of the dam or seep through small cracks in all but the driest periods. One cannot help but have respect for the work ethic of this large, toothy rodent when passing by the results of their nightly efforts.

When Steve Prince, a native New Zealander, realized that his New Hampshire property hosted a family of beavers, he hoped to see a great big sturdy dam, not the unremarkable low one-foot wall he saw winding into the wetland. "I thought we had lazy beavers," he remarks, "until I put my kayak into the marsh and went exploring. The dam may have been low, but it was over sixty feet long!" Beaver dams can be short, spanning a small stream, or long, intercepting the water moving through a wide flowage. The largest dam ever reported was discovered by a researcher scanning aerial photographs of wetlands in Alberta: it was over half a mile long, and is thought to have been started in 1970 and maintained by many generations of beavers.

Where mild weather or consistently high water flow keeps the river or pond from freezing, beavers don't need to build a dam. Instead, they build their lodges alongside the shore or in previously excavated holes in the riverbank. These "bank beavers" swim to their food sources all winter, so they don't need to maintain an underwater cache of twigs for the season. For some reason, these "bank beavers" aren't called by the flowing river; or maybe they know it is pointless and so redirect their energy to more rewarding endeavors such as finding food.

The Beaver's Tail

Ever heard the voice of a beaver, echoing across a pond at twilight? No? Well, there's a reason for that. While most people know that beavers gnaw down trees and build dams, few people know how they communicate — and auditory communication plays only a minor role. Their biggest noise is slapping their tail on the water in warning. Young beavers and their parents will whine at each other, and strangers may hiss, but, as with many mammals, much of their communication is accomplished through scent. In the 1960s, one of the leading researchers of the North American beaver, Dietland Müller-Schwarze, set out to understand the role of the olfactory system in animal communication. Dietland was a student of the Nobel Prize–winning animal behaviorist Konrad Lorenz at the Max Planck Institute for Behavioral Physiology in Germany, at a time when interest in animal communication was on the rise. Many of Lorenz's graduate students were using the new technology of sonographs to characterize songs and other animal vocalizations. But Dietland, knowing that many animals use odor to mark territories and seek mates, thought it would be cool to work with scent. Beavers, like many other species, have specialized glands that produce odoriferous substances. Dietland right away thought he needed a method to visually describe the scent patterns—a "scentograph," if you will. Working with an organic chemist he had met in his apartment building in Freiburg, he learned to use a gas chromatograph to see the patterns produced by the scent glands of several species: red deer in the Black Forest of Germany; chamois, a European wild goat relative with scent glands at the base of the horns; and black-tailed deer in California. Eventually his research group was able to discern the components of animal scents and then use these compounds experimentally. "In Utah, we raised the deer from fawns, and exposed them to different components of their scents and recorded their response."

While conducting these studies in Utah, he saw abandoned beaver dams on steep hillside slopes, but no beavers. He learned that the beavers had been hunted right out of the area. "The locals had just shot them out for target practice. Coming from Europe, where beavers were rare, I couldn't understand this. I felt they were a precious resource—how could they just be shot like that and nobody seemed to mind? I also realized that, since they stay in one place year round, the beaver would be an ideal study animal for the behavioral field tests I wanted to conduct." And so, after establishing himself at State University of New York's College of Environmental Science and Forestry in Syracuse, a twenty-six-year study began.

Eventually Dietland focused his work on Allegany State Park in upstate New York, where he established a long-term research program that involved walking miles along each stream to locate beaver lodges, catching and tagging the occupants, and observing where each family member moved, how long they occupied a site, what they liked to eat, and so on. His research, and that of others, established that beavers in the northeastern United States and Canada build between four and ten dams per mile of stream, on average. Find about a hundred acres of wetland habitat and you are likely to spot at least one or two beaver colonies. With about six beavers per colony each year, that adds up to a lot of large chewing rodents swimming around. However, the population is usually kept in check by limited food supply, harsh winter conditions, and diseases such as bacterial tularemia. Predation, while not common, is also a factor; coyotes, otters, bobcats, mink and even bears have been observed eating the beaver kits, or even an occasional adult (Hardisky 2011). While a beaver may live as long as fifteen years, most live only for a decade or so (Müller-Schwarze 2011). Only about one-third of the newborn kits will survive their first year.

Adolescent beavers leave their natal colony at about two years of age, and may travel just a short distance or as far as nine miles away in search of a tasty patch of poplars or willows along an undammed stretch of stream. These lonesome wanderings expose them to predators, fast-moving cars, difficult terrain, and inadequate food offerings. Males tend to travel the farthest; thus it is no surprise that juvenile male beavers—perhaps like young males of all species—show the lowest survival rate of any age group (Bloomquist and Nielsen 2010).

Dietland's major research questions focused on scent communication among beavers. "Konrad Lorenz always said that it is science if you can accurately predict the events that will happen—that means you understand the system. In August of 1974 we set up our first experiment in the Adirondacks. We tried to lure the beavers out with specific scents, to see how they responded. I knew that trappers had used beavers' own castoreum to lure them into the trap, but they never recorded their behavior." Beavers produce special secretions used in communication through their castor gland, a pair of specialized sacs found between the kidneys and the bladder, releasing into the urethra. They also have anal glands that produce an oily, smelly discharge. Beavers will carry piles of mud from the bottom of the pond, deposit it in a heap near the bank of the pond, then crawl onto the mound, squat, and release the castoreum or the anal gland secretions. By elevating the odoriferous substances, the mound carries the scent farther (Müller-Schwarze 2011).

"So in our first experiment, we made scent mounds just like their own— piles of mud with the scent on top. We sat there in the evening, downwind, with no obstructing vegetation so we could see them. Within a half an hour, they came out and started sniffing, pawing, and then turned around and added their own scent to the mound." The beavers were so drawn to the scent that they did not mind that Dietland brought a large group of on-lookers to spy. This first experiment led to a lifetime of studies, trying to understand how beavers use their scent to communicate. The basic method Dietland used went something like this: First, trap the beavers in a spring-loaded clamshell-type live trap, and then anesthetize them. Next, "milk" the castoreum out of the castor sac or the secretions from the anal gland. ("My assistants and I have probably milked hundreds of beavers in the course of these studies," commented Dietland—a number fit for the *Guinness Book of World Records*, no doubt! "You just rub their tummy a bit, and the casto-reum—which is essentially concentrated urine—just leaks out. For the anal gland, you have to milk it, almost like milking a cow," he explains.) After labeling the vials of secretions with a tagged beaver's identification number, then weighing and measuring the individual, the researchers would put the beaver in the shade and watch it to make sure it recovered as the anesthetic wore off.

Over the years, Dietland and his team established the genealogy of the many colonies of beavers in Allegany State Park. Once the kinship patterns of the tagged beavers were established, Dietland and his colleague, Lixing Sun, experimented by creating scent mounds from related and unrelated beavers, and found that if the scent was left by an unrelated individual, the resident beavers would rip the scent mound apart and build a new one with their own scent in its place. They found that the greater the density of bea-ver colonies, the more scent mounds the beavers created—like neighbors in a densely packed housing development putting up more fences to protect their own territory from intruders. If the artificially created scent mound carried the secretions of a relative—even one they had never met—the bea-vers would not behave as aggressively. The secretions from the anal gland seem to compose what Dietland calls the "olfactory identity card" (Müller-Schwarze 2011). Over the years, research established that the composition of the castoreum is largely determined by the beaver's food resources, thus communicating by scent if the beaver is well nourished. The anal gland se-cretions seem to be more genetically determined, thus likely to be the better indicator of kinship (Müller-Schwarze 2011).

This work entailed countless hours of quiet surveillance by several ob-servers sitting downwind in a spot with a good view of the scent mounds, lodge, and pool. Sometimes an observer was so quiet that unusual sightings did occur: on one occasion, Dietland's wife and long-time research partner, Christine, was sitting in one area with just her head showing. "She was very intent on making observations," Dietland explains, "but something made her look up. She glanced over her shoulder and saw a black bear rearing behind her. Her heart was really racing. This was before cell phones—there was no way to get any help. She was very scared, and sat quite still as the bear moved in a big circle around her. The bear seemed to be examining her, trying to figure out what it was he was looking at, in order to decide what she was and whether she was a threat. Finally, he left." Just another day in the life of a field biologist—good data, and good stories.

Through the years, Dietland and Christine saw many changes in the watersheds as the beavers moved around. Dietland describes the characteristic cycle that follows the return of the beaver, bringing all kinds of other wild creatures in its wake: "First, the trees will die, and that provides snags for all kinds of birds, and places to live. Then you have animals moving in: amphibians move into the shallow water, newts and tadpoles. If there is a beaver meadow adjacent to the pond, you get lots of butterflies; in fact, the beaver meadows at our field site were a favorite spot for the entomologists in our biology department. The cycling of the vegetation is particularly interesting—they might cut an acre of aspen, then the beavers leave. So the aspens sprout again, but the beavers don't come back. Even though to us those sprouts look nice and tasty, it takes close to eight years for them to come back and start cutting the aspens again. We know from other studies that those aspen sprouts are heavily defended, so if the beavers take a quick bite, they find it doesn't taste good. You can see places where the beaver has sampled the tree and rejected it."

Many woody plants defend themselves by producing noxious substances that should nauseate or kill any insect, rodent, or other creature that chews on them. But when ingested by beavers, these chemicals—which include benzyl alcohol from aspens and poplars, and salicylaldehyde from willows (the pain reliever found in aspirin)—are routed to the castor sacs. Specialized structures in these sacs lock up these potentially poisonous substances, keeping them away from any sensitive organs. When the beaver secretes castoreum, it is getting rid of harmful poisons as well as using them to warn intruders. Thus, the beavers recycle the chemical weaponry of plants to defend their own territory.

The comical, cartoon image of the "busy beaver" is not without any basis in reality. Beavers never stop chewing, carrying, swimming, and building. In just one year, a colony of beavers can cut more than a ton of wood, all taken from within about three hundred feet around their pond, which equates to approximately 40% of woody plant biomass removed in half a dozen years (Rosell et al. 2005).

A meander through the area around a beaver pond reveals that many trees are chewed only partway through, and it is easy to think that perhaps the beaver got tired or distracted from its task. But it is not all mindless gnawing—they do have a method to their madness. This partial gnawing, or "girdling," is a time-tested method of forest management. The beavers do it to kill unpalatable trees such as hemlocks or other conifers such as fir, hemlock, and larch (Müller-Schwarze and Sun 2003). When these trees die and fall, the sunlit, open spot is available for more desirable and fast-growing tree species, such as aspen, willow, poplar, alder, ash, hazelnut, and black cherry, and, to a lesser extent, red maple, oak, mountain ash, and birch (Müller-Schwarze and Sun 2003). In drier areas further from the wetland edge, the beaver's attempts at forestry may backfire, and the light gaps it creates may allow inedible, light-loving species such as pine and spruce seedlings to thrive (Rosell et al. 2005).

Survival is also jeopardized by the beavers' everyday work activity. Given our understanding of their excellent lumberjack skills, it is surprising to learn that it is not uncommon for beavers to be killed by the tree they are felling. As a child, Dan Houghton, a Connecticut businessman and avid conservationist, witnessed this firsthand in the wetland behind his childhood summer home in Spofford, New Hampshire. "I was seven or eight years old, and my brother and I were visiting my grandmother during spring vacation. Craig and I were in the habit of traveling down the path every day to see what was going on in the beaver pond. We came around a corner and saw this big beaver with its tail pinned by the trunk of a tree." Apparently, the beaver had attempted to cut a poplar, but the tree had hemmed up in the forest canopy, so it didn't fall the way the poor animal had anticipated it would. "I was surprised at how round and how big he was, and I was struck at how bright orange and chiseled his teeth were. We ran back to the house and got the chainsaw. My brother cut the tree in two places and pushed the trunk off of him, and he scampered right off. We checked the next day, thinking if he was badly injured we might find him dead, but we never saw him, so I guess he was okay!"

This incident brings to mind a quote from naturalist Ann Zwinger (1970): "A beaver does not, as legend would have it, know which direction the tree will fall when he cuts it, but counts on alacrity to make up for lack of engineering expertise." So much for quick thinking, in this case.

The Pool

As a dam becomes more substantial, the stream's reach extends over the adjacent low-lying areas and the resulting pond behind the dam gets deeper and wider. As the water builds up, its flow diminishes and its inky darkness intensifies. Soon, woody plants whose roots had previously known only brief dunkings are sitting in stagnant water, desperate for air. Within a few hours, cells in the roots are sending alarm signals as their ability to produce energy is compromised by the decline in oxygen. Lacking oxygen, plant respiration switches to an anaerobic (no-oxygen) pathway, and the plant poisons its own cells with the acid by-products. Leaves turn yellow and start to drop. Like an overwatered houseplant, roots drown even though leaves have plenty of air (see chap. 1, box 2, for more details). Within a few months, most species of woody plants are dead.

Although a number of woody wetland plants have structures and mechanisms necessary to survive oxygen deprivation in the deep water behind a beaver dam, only a few can sit in water for an entire growing season (McIninch and Biggs 1993). Trees at the edges of wet areas are always dealing with low-oxygen stress induced by wet conditions, which explains why the first vivid reds of autumn are displayed in the leaves of the red maples living in or near wetland edges. The stress of waterlogging reduces photosynthesis, so the maples have less food energy to maintain themselves. End result? The green chlorophyll in the leaves breaks down early, revealing the glorious reds of the anthocyanin and lycopene pigments.

Even for the trees left standing, the impact of flooding is so intense that ecologists can detect the year the dam was built and the year that beavers abandoned it by looking at tree rings: a sudden narrowing is considered a "pointer year," with needle-thin rings indicating reduced growth resulting from the inability to obtain oxygen. Wider rings are a tree's sighs of relief, big breaths of oxygen made possible again by a drop in water levels (Little, Guntenspergen, and Allen 2012).

In the first year or two of flooding, the beaver pond's glassy surface will be broken only by tree trunks, shrubby branches, and tips of grasses peeking through, gasping for air. The deep green of winterberry (*Ilex verticillata*) and the more delicate lemony-green leaves of meadowsweet (*Spiraea alba*) may persist for a while, although no new growth is added. Sitting in several feet of water, the shorter grasses, sedges, and wildflowers will succumb quickly to oxygen deprivation. At the same time, organic matter in the water produces dark-brown tannic acids that turn the water the color of a deeply steeped cup of tea. The result for the plants is starvation because, with so little light penetrating the darkened waters, they are unable to photosynthesize to make food.

Over a number of years, the flooded area is colonized by aquatic plants adapted to deeper waters, such as water lilies (*Nymphaea odorata*) and pickerelweed (*Pontederia cordata*). The lack of oxygen that killed the earlier swamp denizens is no challenge for emergent cattails (*Typha* spp.), arrowhead (*Sagittaria* spp.), bur-reed (*Sparganium* spp.), bulrushes (*Scirpus* spp.), or spike rushes (*Eleocharis* spp.). All these plants employ spongy, air-filled leaf spaces to bring oxygen to their roots through specialized cells called aerenchyma (see chap. 1, box 2).

When the area upstream of the dam is a bog or a fen, the organic, spongy mat of the bog just floats up as the water levels increase. Plants experience the same depth of water, so the floating ecosystem remains unchanged.

If They Build It, Others Will Come

The beavers' efforts to deter others from invading their space works only on a few species, however, as the pond they have created provide perfect conditions for many types of creatures. Once the beavers have built the dam, the rate of water flow decreases, sediments are trapped, and gravel, leaves, sticks, and other debris are deposited behind the dam. This material creates fine habitat for a number of invertebrates—particularly the predaceous larvae of dragonflies and their more delicate cousins, the damselflies, as well as tubeworms, midges, and freshwater clams (Müller-Schwarze and Sun 2003). Not all aquatic insects enjoy these conditions, however. The exposed gravel favored by stoneflies disappears under the fine sediment, and as many as seven different taxonomic groups that used the free-flowing river habitat will decline (Naiman, Johnston, and Kelley 1988). Happily, included in the list of negatively affected species are black flies, whose gruesome cutting mouthparts inflict such pain on exposed human flesh each spring—yet another reason to be grateful for beavers. In most situations, a newly created beaver pool will result in an increase in diversity and abundance of the types of invertebrates that benefit from the muddy bottom and decaying wood (Müller-Schwarze 2011).

All these insects, from their larval, gilled stages through various juvenile instars to their adult bodies, create an excellent cafeteria for fish. These creatures, and the refuges created by submerged logs and sticks, lead to greater numbers of warm-water fish species—particularly mud minnows, northern pike and smallmouth bass. Chain pickerel, sunfish, shiners, fatheads, brook trout, largemouth bass, and many others all grow bigger in beaver ponds. Not only do they find enough food, but they have places to hide; and the slower velocity of the water in the pool means that they can hunt in a leisurely manner, expending less energy as they nose around in the shallows or hide from the heat in the depths (Pollock, Heim, and Werner 2003). While the dams may impede fish migration, spring flows are often high enough to allow trout and pike to leap over the barrier.

Because the water temperature in the pond is higher than it was in the free-running river, some cold-water species such as trout may move elsewhere (Müller-Schwarze and Sun 2003). In tidal areas, anadromous fish such as salmon, which migrate from the ocean into freshwater to spawn, may be more abundant in beaver ponds. Evidence suggests that fish enjoy protection from predators such as great blue herons, whose prey-catching prowess is compromised by deep waters (W. G. Hood 2012).

Deep, open water is critical habitat for many amphibians and reptiles, too. Many turtle species prefer the slower water, mucky bottom, and abundant prey—including tadpoles and insect larvae—of the beaver pool, compared to the undammed stream that preceded it (Rosell et al. 2005). The early spring chorus of wood frogs, sounding for all the world like a flock of nasal ducks, beckons explorers to the beaver pond on a warm spring day. These and other vernal pool species (see chap. 6) will use the shallow, plant-packed edges of the pond for breeding, as these areas are unlikely to be visited by predatory fish. At night, the male spring peepers begin their peep-peep-peep serenade, wooing the females who have hopped through the forest to find a mate and a place to start a family. Yellow-spotted salamanders will also visit the pond to perform their own spring mating rituals.

Standing in or near the deep water of the pond are the tall gray reminders of the forest that once existed. These trees, killed by the beaver flooding, are now dead snags, full of hidey-holes for the nests of chickadees, flycatchers, wood ducks, hooded mergansers, and owls. Insects that feed on the dead and dying wood attract woodpeckers, who chip away at the wood to get to the insects. Pileated woodpeckers, the largest of the North American woodpeckers, will carve out rectangular holes in the wood—sometimes a whole hand and half deep. These holes may be further excavated by brown creepers, chickadees, or others who, after scattering the excavated wood chips some distance away, will then line the little cavern with cattail down, moss, feathers, hair, or insect cocoons to build a cozy nest for their offspring.

Where a branch has fallen from a snag, or if the tree is rotting from the inside, a larger hole may form and can become the summer home for other, bigger birds, such as screech owls, wood ducks, or hooded mergansers. Hooded mergansers, or hoodies, are small diving ducks who use the serrated edges of their narrow bills to snag fish from the pond's murky water. The male hoody attracts the female with a rounded, black-edged, white head crest, which he shows to advantage with impressive head-bobbing and neck-bending moves. After he gets his gal, she flies into the nest cavity and lays ten to twelve eggs. These eggs hatch into downy, striped ducklings, ready to leap out of the tree cavity as far as fifty feet down into the water, less than twenty-four hours after hatching. If the mother duck was not lucky enough to find a tree cavity near water, the little fluff balls have to follow her for more than a mile to find open water, creating a fine opportunity for predators to catch a tasty snack. When the beaver wetland is close to a number of other beaver-created ponds and adjacent wetlands, waterfowl like mergansers and wood ducks benefit from reduced travel time for their offspring, which translates into higher survival rates and larger duck populations.

A large group of standing trees in a beaver pond or reservoir may become a choice nesting site for the local great blue heron (*Ardea herodias*) population. The graceful herons, looking more light gray than blue, are the tall birds we see standing patiently at the edges of flowing streams or shallow ponds, seeking frogs and fish to spear with their long beaks. The great blue heron nests in a colony called a rookery, each of which may contain anywhere from five to five hundred nests. Seeking protection from predators, herons will choose trees in standing water or on islands and may share the rookery with their elegant white cousins, the great egret (*Ardea alba*), or their short, secretive relatives, the black-crowned night herons (*Nycticorax nycticorax*). The rookery, with pairs of long-necked herons sitting atop huge, messy nests way up high in the tree, create a Dr. Seuss–inspired landscape. The herons sit on their clutches of smooth, light-greeny-blue eggs for about four weeks, until as many as seven hungry, skinny nestlings hatch out at each nest. Within six weeks, the chicks transform first to quarreling, gangly teenagers, feathers sticking out every which way, then to fledgling subadults, launching awkwardly out of the nest. Many heron rookeries last for decades (Spurr 2003), but in some cases, the big birds produce so much nitrogen-laden, caustic excrement that the few surviving trees—already struggling for oxygen in the standing water—slowly die, and the birds have to find new nesting trees nearby or relocate the whole rookery in some other beaver pond or reservoir.

Fur-bearing wildlife also use the beaver pond. Muskrats, like beavers, are swimming rodents, but they are smaller and have a long slender tail; they enjoy the cool swimming opportunities and the dense cattails found in the beaver pond. They also build lodges (smaller versions of the beaver homes, made of aquatic plants and mud). Muskrats feast on cattails, and can turn a solid stand of dense cattails into a nice mosaic of open water and remnant plants in short order. Otters, too, benefit from the increased open water of the beaver pond, with its offerings of fine-quality otter fare such as golden shiners, chubs, and freshwater mussels.

A Big Dam Difference

Flora and fauna alone are not the only beneficiaries of the beavers' endeavors. Considering how the stream has now been partially converted into a pond, we can appreciate the physical impacts of the beavers' handiwork on the watershed. By capturing and holding large amounts of water, the dam transforms the flow, allowing only small amounts of water over or through the "leaky" dam. Without the dam, the stream may have held flowing water just in springtime—particularly if it is in the upper headwaters section of the watershed. With the beaver impoundment, the dam retains much of the water that would have flowed downstream right after snowmelt; it continues to release streamflow through most of the growing season, reducing any drought effects (Rosell et al. 2005). Depending on the soil type and slope, the impoundment will result in a large wetted area around it, from which water can slowly infiltrate into the aquifer below, thus maintaining groundwater supplies (Naiman, Johnston, and Kelley 1998; Rosell et al. 2005).

The profound transformational power of beaver colonies was illustrated by a study conducted in northern Minnesota (Naiman, Johnston, and Kelley 1988). Comparing aerial photographs of the same 450 square kilometers (174 sq. miles) of landscape, the authors found 17 dams in 1940 and 835 in 1986. The newly dammed landscape in the more recent photos displayed a complex pattern of thirty-two different vegetation types, reflecting ponds of different age and depth, shorelines with varying numbers of felled trees, and abandoned impoundments in different stages of regrowth and in different physical settings.

In some mountainous areas, beavers build dams in isolated wet basins that have no streams (G. Hood and Bayley 2008). In these areas, the beaver dams capture overland flow, also called runoff, rather than streamflow, transforming what would have been a merely a damp depression into a good-size pond. The dam and the pond hold the water that would have otherwise simply flowed downhill or evaporated. In dry areas or in times of drought, this makes all the difference—retained water is available to percolate into the groundwater, keeping the area moist. Research in Alberta, Canada, found that, as the number of beaver lodges (and the accompanying dams) increased, the area of open water increased, regardless of the amount of precipitation. In other words, beavers were more important than rain and snow at maintaining open water and wetlands (G. Hood and Bayley 2008). Even in a wet year, if there aren't many beavers around, the land will be considerably drier (G. Hood and Bayley 2008). In fact, the ability of beaver dams to store spring runoff is so important that at least one state, Oregon, has shelved plans for concrete dam construction in favor of beaver reintroduction (Groc 2010). As climate change warms the planet and alters rainfall patterns, causing some dry areas to become even drier—the role of beavers as waterkeepers on the landscape becomes even more essential.

Much of the eastern United States and Canada is expected to become wetter, not drier, as a result of climate change, with more precipitation coming in large, often violent, storms. Some dams may reduce flooding and downstream erosion during a large storm because they hold the water back and slow down the velocity of the stream. This allows the stream to modulate the effects of climate extremes on the watershed. The water will be shadier, and cooler, along the beaver impoundment, and thus more able to withstand rapid evaporation associated with drought conditions. The stream, forced by the dam to meander over a larger area of sediment deposits, creates a complex of shallow-gradient, lower-energy pools and meadows, better able to absorb the input of large storms.

But if the flow from upstream of the dam becomes too great, watch out! The sudden, forceful flow after a beaver dam failure can release a wall of water large enough to undermine and wash out small bridges and paved roadways over the stream (Rosell et al. 2005). The muck and sediments that accumulate behind the dam will have impacts on the streambed for years to come. Massive amounts — anywhere from 50 to 8,500 cubic yards, enough to fill between nine and nine hundred dump trucks — of these materials build up, raising the elevation of the pond's bottom and reducing the water depth, eventually spreading out over the valley floor upstream. Plants colonize the muck, slowing down the water and trapping more sediment. At some point, the sheer volume and weight of the muck and mud would either fill in the pond completely or cause the dam to blow out; but, in most cases, the beavers will abandon the dam long before this happens (Pollock, Heim, and Werner 2003). However, the impacts of these structural changes will remain after the dam and its creators have left — beaver dams will actually create permanent alterations in the shape, slope and ground elevation of the river valley. What would have been a narrow streambed chiseled into the landscape and running down a steep gradient becomes a wider, sediment-covered valley, as the dams force the stream to meander. Instead of a steep downhill rush, the flow way descends in a stairstep pattern from dam to dam (Naiman, Johnston, and Kelley 1998; Burchsted et al. 2010). In this way, beaver dams can divide the river into sections, like cement dams created for industry, flood control, or hydropower. Unlike these human-made obstructions, though, beaver dams are not permanent, and their leakiness and short stature allow water and creatures to flow, climb, or jump through, around, or over them (Burchsted et al. 2010).

Some negative impacts can occur because of the dam. The stagnant water in the pond quickly loses oxygen as organic matter decomposes, because naturally occurring bacteria use the oxygen to break down the organic matter. When the oxygen level drops, essential nutrients—particularly phosphorus—are released from the sediments under the water. This phosphorus can lead to an algae bloom if the pond is large enough, and can stimulate growth of cattails or the invasive reed *Phragmites australis*, both of which grow most happily when fertilized this way. A sedge meadow or shrubby wetland once dammed by beavers may thus be invaded by these aggressive plants, which will often persist even after the beavers have left, resulting in a less diverse monoculture of tall, crowded plants. This fertilization process is sometimes called internal eutrophication, similar to the eutrophication of lakes that occurs when too many nutrients wash into it from lawn chemicals, soil erosion, pavement runoff, or other sources.

The Cycle: Dam-Eat-Move

Beavers do not stay at a single spot forever; the site may be occupied for as many as forty-five years, or for just a few seasons. When the beavers run out of their favorite foods, they move on. According to ecologist Tom Wessels (1997), when the beavers start eating hemlock, their poverty food, you know they will abandon the area soon.

Without the nightly maintenance from their toothy creators, the dams will slowly wash away. As the water levels recede, a sequence of events unfolds. The rich sediments that accumulated behind the dam are revealed as a shallow bowl of dark muck. This open mudflat will be colonized by small plants, which may have lain dormant in the seed bank, or were carried there on wind gusts, water flows, or animal toes. Small sedges such as silvery sedge (*Carex canescens*), swamp candles (*Lysimachia terrestris*), and marsh Saint-John's-wort (*Triadenum virginicum*) appear within a few years. In the drier areas along the shoreline and on hummocks, woody shrubs such as speckled alder and northern arrowwood (*Viburnum dentatum*) may start to grow. A marshy field, like the Lord's Meadow at the start of our story, is born—again.

This successional pathway takes different twists and turns depending on the specifics of the location. In wetter locations, the small sedges and delicate herbs are replaced within a decade by the taller, more competitive sedges and grasses such as tussock sedge (*Carex stricta*) or bluejoint grass (*Calamagrostis canadensis*), and small shrubs such as meadowsweet.

Some researchers have found that, in many cases, this sedge meadow will be the stable "end game" for the area (Little, Guntenspergen, and Allen 2012), when the resulting water levels are fairly shallow (less than ankle-deep). In other areas, as the breakdown of the dam allows the water to seek a lower level, shrubs such as speckled alder (*Alnus incana* subsp. *rugosa*), high-bush blueberry (*Vaccinium corymbosum*), wild raisin (*Viburnum cassinoides*), maleberry (*Lyonia ligustrina*), and others will invade as the area dries out. If the beavers leave the area alone for a good forty years and water levels drop even lower, the area may become dry enough to support a forested wetland of red maple and gray birch, or a coniferous swamp dominated by spruce, fir, and tamarack (Little, Guntenspergen, and Allen 2012).

Even in places where the water is shallow, if the natural water chemistry brings acidic groundwater and supports a layer of undecomposed plant matter (peat), a sedge meadow may never be established; instead, a shrub fen grows, full of thickets of ankle-catching leatherleaf (*Chamaedaphne calyculata*), sweet gale (*Myrica gale*), and skinny wiregrass sedge (*Carex lasiocarpa*) sprouting through thick mats of sphagnum moss (*Sphagnum* spp.) (Little, Guntenspergen, and Allen 2012).

When the beaver pond persists for decades, however, the soil becomes so oxygen-depleted for so long that certain microorganisms cannot survive. Black spruce (*Picea mariana*), for example, needs mycorrhizal fungi to germinate, but the fungi can't tolerate wet conditions. After a prolonged dunking, the required fungi have all been killed, and can be found in the adjacent uplands but nowhere else. Curiously, the only way the spores (primitive versions of seeds) of those essential fungi are able to travel to the now drier wetland area is in the food and feces of red-backed voles (*Myodes rutilus*). These short-tailed rodents will venture back and forth from the upland to the marsh, carrying the fungi on their food and feet. But meadow voles (*Microtus pennsylvanicus*), who do not like to journey into the upland, tend to dominate the wetland and their aggressive behavior discourages visits from their red-backed cousins. Thus, even though the water levels have dropped and conditions seem perfect for reestablishment of the forest, many years may pass before the black spruce seeds can germinate. In some northern areas of the beaver's range, the wet meadow phase can last seventy years or more; it takes this long for enough of the essential fungi to hitch a ride on the rare visits of the unwelcomed red-backed voles (Moore 1999).

The presence of a wet forest along a stream today indicates that the beavers may have been absent from the site for almost half a century; so, these forested wetlands may be living remnants of the time when the land was beaver-free. Similarly, according to Tom Wessels in *Reading the Forested Landscape*, many large standing-dead trees in a wetland attest to a time when the area was beaver-less long enough for trees to grow back and grow big, before beaver reintroduction drowned them out.

As the beaver population expands, these forests will see fewer opportunities to establish. Trapping of beavers, which led to their near extinction in the 1800s, is on the decline. Most states ban the notoriously cruel toothed-leghold traps now, and some states such as Massachusetts only allow box or cage-type traps. Without trapping, wetland areas experience a two- to threefold increase in beavers (McCall et al. 1996). Hunting of beavers is allowed in some states. However, most people agree that, short of some huge new demand for beaver products, trapping or hunting are short-term solutions anyway—as long as the habitat exists, new beavers will move in to take advantage of the resources left behind when the previous occupants were forced to relocate. Trapping can actually increase the beaver population: taking advantage of these newly available resources, a well-fed female beaver in an area of few beavers will produce 33% more kits than females in a more densely populated area (Hardisky 2011).

As a result, suitable wetlands in the northeastern and midwestern states are chock full of beavers. The North American population of beavers is estimated to be ten million to fifteen million, from a low of one hundred thousand in 1900. Coyotes—the main predator of beavers—as well as disease and high juvenile mortality are doing their job to keep the number down; nevertheless, widespread beaver activity has left many landowners contending with flooded driveways, basements, septic systems, and lawns.

Troublesome beaver dams have led to an ever-escalating battle between homeowners and these marvelous rodent landscapers all across the country. Angry landowners, desperate road agents, concerned fish and game officers, and paid professionals have put out traps to catch the critters, or gasoline-soaked bags and coyote-scented rags to deter them, only to have the materials—including the steel traps—sometimes incorporated into the dams as building materials! A day spent deconstructing the dam by hand, bulldozer, chainsaw, or dynamite will be followed by a night of frenzied beaver activity; next morning's light will reveal the dam to be well on its way to completion. Homeowners like Scott Monette of South Royalston, Massachusetts, watch their streams and ponds closely. "One year they dammed the culvert so well, the driveway was flooded—you couldn't get in or out. Every day, I'd take down the dam; every night they'd rebuild it. I'd throw rocks at them, I'd have the dogs chase them—they just didn't leave. I like watching them—they're cool animals—but this just couldn't work." Fed up, Scott took an ecological approach to the problem: "Finally I just went and chopped down all the aspen and willow around the pond, took away their food. They just moved upstream. There's plenty of habitat around here, but at least my road isn't underwater!"

This is just one of a number of successful tactics that have been devised to effectively prevent beavers from damming a stream area, bringing about a truce in the battle of the beaver. Other homeowners have picked up and moved a dam upstream where flooding wouldn't bother anyone. A fence can also be constructed directly in the stream, around the upstream side of the culvert; beavers will build a dam around the fence instead of on the culvert, or will avoid the area altogether. Another approach is the beaver excluder, a tube-and-fence contraption installed in a culvert that can't be dammed but which does allow water to flow. There is also the beaver leveler, a pipe installed through the dam with a right-angle upturn to keep the water level the same at all times. It has to do this relatively quietly, since the sound of running water will bring in the beavers and incite them to dam it all up. Where prize trees and shrubs are being chewed, the base of each tree or shrub needs to be wrapped with galvanized steel mesh or painted with a mix of sand and latex paint.

Recognizing the many benefits that beavers bring to the landscape, several writers and organizers have suggested that humans should view beavers as essential partners in the quest for sustainability rather than as threats to the human-dominated landscape. Their work with mud and sticks, building, damming, chewing, eating, and then moving on, creates a mosaic of different vegetation types along the valley: deep, cool ponds and pickerelweed patches, alder thickets and mudflats, silvery sedge meadows and dark spruce forests, red maple glens, bluejoint grass jungles, and dead tree stands full of heron rookeries. These profound alterations in the landscape also serve to protect water supplies, stabilize streambanks, absorb large swings in climate and precipitation, and create cool, damp places for fish to flourish, turtles to hunt, and wood ducks to call home. Beavers, as much artists as engineers, practice their craft in forgotten corners of subdivisions, cities, parks, and protected lands, bringing excitement, enjoyment, and ecosystem health to the landscape.

For wildlife watchers like landowner Dan Houghton, who rescued the beaver trapped by its own treefall, the beavers are something special. "Walking down to that beaver dam was the first thing we'd do when we got up in the morning. We'd sleep out there, and we'd hear the beavers slapping away. It has always been a special place to us. My grandfather, my father, my cousins, my brother, and I always walked the path to the stream and visited the beaver dam. My dad didn't see the beavers as flooding the property and ruining the land. He just liked them. On the night when my dad passed away, we saw a double rainbow that seemed to end at the beaver dam, and that seemed just right."

CHAPTER 4

Stuck in the Muck: Bogs and Fens

The bog is a strange and dangerous place, neither land nor water—
a desolate landscape with neither roads, nor paths, nor fixed points,
just a bottomless deep waiting to engulf the trespasser. Take one wrong
step while jumping from tuft to tuft and you will perish. The bog is alluring
and seductive and inhabited by strange creatures. The dreamlike mist
across the black deep has fuelled myths and ballads about ladies of the
bog and will-o'-the wisps and gorgeous elfmaidens. He who went there
at night on the grounds of desire was sure to be lost.
—*Treacherous and Alluring Bogs*, wall display at the Moesgaard Museum,
Højbjerg, Denmark, quoted by Stuart McLean, "Bodies from the Bog"

Clinging to the uppermost portion of the scraggly spruce trunk, peering over
a canopy of similarly skinny spires, I wondered how long it would take someone to find me if I fell. I still couldn't see any landmarks — only treetops and
a low-lying, gray mist. I was lost. Lost in the peatlands of western Maine.
What kind of an idiot ventures into a thousand-acre area of undeveloped
forest and bog armed only with a backpack of research gear and a ham sand-
wich? Wouldn't a *compass* and a *map* have been useful additions to the field
kit? The day was pre–cell phone, pre-GPS, but those modern conveniences

may not have helped anyway, it was too remote for cell towers or even decent satellite coverage. I was lost.

I hadn't needed any navigational aids before. We had "flagged" the path early in May, tying pink ribbons of plastic flagging at regular intervals along a line due north from our parking spot along the logging road into Johnson Bog. The bright pink ribbons were impossible to miss, or so I thought; standing at any one of them, you could see at least two in a row in either direction, providing linear navigation through the thick shrubs and scrawny trees. Our path bisected the concentric rings of plant zones that are so common in true bogs. The vegetation arranges itself in response to the water levels, water acidity, and the depth of the peat that arose during formation. Around the very edge, there is a shrubby, wet zone called a lagg, where surface water and groundwater from the surrounding upland collects, making it too wet for trees. Here grows a tangle of tall shrubs, such as highbush blueberry (*Vaccinium corymbosum*), winterberry (*Ilex verticillata*), and wild raisin (*Viburnum cassinoides*), intermixed with some sedges. The lagg zone then transitions to a forested zone of conifers, notably black spruce (*Picea mariana*). A zone of short shrubs is next encountered, such as leatherleaf (*Chamaedaphne calyculata*), bog laurel (*Kalmia polifolia*), and Labrador tea (*Rhododendron groenlandicum*). Closer to the central open bog, the trees and shrubs became shorter and sparser, while the sphagnum moss (*Sphagnum* spp.), skinny sedges, and leatherleaf became dominant. At the center a lush, colorful carpet of sphagnum, dotted with the puffy heads of cottongrass (*Eriophorum angustifolium*), bright red fruits of small cranberries (*Vaccinium oxycoccos*), and hungry purple pitcher plants (*Sarracenia purpurea*), surrounds an open pool of water.

After passing through these zones along the trail, the entire two hundred acres of the bog mat and pond is clearly visible, its low vegetation making it easy to find the arrays of drift fences and pit traps we had constructed there. My job as a field assistant was to walk one mile along the trail and take the tops off the cylinder of joined, double-high coffee cans we had sunk into the ground in several locations along the path, ending in the open bog. Two days later, I'd return and identify all the amphibians and small mammals that had fallen into the cans. By mid-June, I'd done this three times, without any navigational challenges.

But now, all the deciduous shrubs and small trees—red maple, black ash,

Box 3. Bog vs. Fen

When is a bog a "true bog"?

- There is no streamflow into or out of the wetland.
- Sphagnum moss and heath plants such as Labrador tea and black crowberry are common.
- Acidity is very high (pH < 4).
- You are in a northern region (latitude >43 degrees N).

When is it a fen?

- A stream runs into or out of the wetland.
- Sedges, bladderworts, buckbeans, and three-leaved Solomon's seal are found among the vegetation; sphagnum moss may also be common.
- The spongy soil below you is derived mostly from sedges and woody plants.
- Acidity is moderate (pH > 4).

mountain holly, huckleberry, highbush blueberry, wild raisin — in the border zones had unfurled their leaves to their fullest extent, obscuring my pink-ribboned path. Even missing that straight path by a few degrees could leave me walking for miles, lost in the tree and shrub zone, before I'd hit one of the logging roads or the river. Treacherous indeed.

All I needed to do was catch a glimpse of either the bog or the road and I could find my way. So I climbed a tree. The trees neither were very tall, nor did they sport sturdy branches. Shimmying up the swaying tree trunk, past bristly needles and poking branch stubs, I questioned the wisdom of this strategy. About the time I was ready to give up and jump (or fall) down, the low-lying fog lifted, and there it lay: the open bog, with the pond in the middle. I was found.

The word *bog* elicits eerie images of ghostly figures in long white dresses fleetingly glimpsed, insubstantial ground waiting to ensnare unsuspecting feet, red glaring eyes of predators peering from the dark spruce forest. But peatlands, of which true bogs are but one type, are actually enchanting places full of dwarfed trees, richly hued shrubs, and finely textured mosses. Many people call any wetland a bog, particularly if it has a squishy surface and is covered with mosses and low shrubs. Most of these wetlands are not true bogs, but more likely a kind of wetland called a poor fen, or some other wetland type entirely (see box 3).

Both bogs and fens are spongy places, filled with years of accumulated peat—*peatlands* is the generic term. The bogs of northern Minnesota, the moors of Scotland, the mires of Sweden, the muskegs of Hudson Bay are all examples of peatlands. Again, peat is partially decomposed plant matter. When plants grow, they conjure leaves, stems, and roots out of carbon dioxide, sunlight, and water. Every year, even healthy plants lose some of themselves—flowers drop, leaves fall, branches break, roots rot, stems die back. Where it is dry and there is plenty of oxygen, the tiny bacteria, fungi and other small creatures that power the earth's recycling system break these parts down into soil and carbon dioxide gas. In wetlands—in particular, stagnant, cold ones like bogs—these decomposers are choked by lack of oxygen and chilled by low temperatures. They can't keep up with the accumulation of plant parts. Peat builds up, but not quickly; it takes anywhere from one hundred to nine hundred years to create one foot of peat (Crum 1992). The wetter and colder it is, the deeper the peat.

Natural Preservatives

Many wetlands have some peat. True bogs have particularly deep peat deposits—as deep as twenty or thirty feet, sometimes more. In addition to the regionally wet, cold conditions that inhibit decomposition and allow peat to accumulate, bogs also have another weapon to wield at the decomposers: acid. Water in most true bogs has a pH not much higher than 4, the same pH as tomatoes, wine, or beer, although not nearly as tasty.

Deep layers of acidic peat are so inhospitable to the bacteria, fungi, and worms which break down living materials, that animal remains—including humans—have been found with hair, eyelashes, and clothing still intact after hundreds of years. Even stomach contents are recognizable, revealing the interred victim's last meal. Although bog waters dissolve calcium-rich bones, they turn human skin and hair wild shades of russet brown (literally tanning the hide), so detailed features of flesh are clear: the shape of a nose, the arch of a foot, pores in the skin of a cheek.

Famous stories of these "bog people" are abundant, but a perennial favorite is Clonycavan Man, a young man whose body was found in a bog in the central Irish county of Meath in 2003. As archaeological journalists Jarrett Lobell and Samir Patel tell it, Clonycavan Man lived over 2,200 years ago, during the time when the Celtic people prevailed (Lobell and Patel 2010). Analysis of his hair, which he styled into a mohawk using plant resins, revealed that he ate a rich vegetable and meat diet, luxuries not available to everyone. On his biceps he wore a leather armband adorned with a stylish brass amulet. His body showed no evidence of the wear and tear of a working man's life, and he had been very healthy when he died—if you overlook the multitude of axe wounds to the head and torso. All of this suggests that Clonycavan Man was a rich man, most likely a failed king, killed and sacrificed to the gods in response to a drought or other calamity. "The bodies served as offerings to the goddess of the land to whom the king was wed in his inauguration ceremony. . . . The multiple injuries may reflect the belief that the goddess was not only one of the land and fertility, but also of sovereignty, war, and death," the authors surmise (Lobell and Patel 2010). One fatal wound required per divine realm, apparently. Burial in the bog may have also served as a signal to neighboring clans: Keep away—our goddess is Not Nice, and she will avenge any territorial transgressions.

While I knew that this method of bog burial was common only in Europe, it was always on my mind in the Maine peat bogs, particularly while digging into the peat to install the sunken drift fences and coffee-can pit traps for the project at Johnson Bog. Reaching down with trepidation into the two-foot-deep hole in the peat to extract a decidedly arm-like obstacle, I tried not to envision the orangey, contorted faces of the bog people I had seen in *National Geographic*. Tension turned to relief as I pulled up a tree branch— harmless enough, albeit sometimes bearing a ghoulish resemblance to a preserved human limb in heft and hue. A layer of trees and shrubs buried in the peat like this tells us that the area was once dry enough to support tree growth. This may seem less intriguing than finding grim evidence of foul play, but to a small, isolated team of ecologists working in a remote area, the proof that nonhuman natural forces are at work (rather than angry death goddesses or mutilation-happy villagers) is reassuring.

But it is more than just bodies and tree branches that get preserved in the cold, acidic bog waters. The peat also preserves the pollen that blows in from surrounding lands. As the peat builds up over time, the pollen layers become a record of what kinds of plants used to grow in the area. Combining methods that tell the age of the peat at different depths with identification of pollen at each depth provides a historical record of how the land has transformed, largely in response to changing climates. Pollen of trees such as cedar, spruce, and tamarack were deposited in wet areas during cold times, especially during the last glaciation in North America. As the climate warmed and the glaciers melted, oak became more common in the mid-Atlantic states, and moose, elk, caribou, and musk ox migrated north with the spruce and fir. Some areas show more beech and hemlock in the depths of the peat, indicating wetter conditions some six thousand years ago (Crum 1992). The preservative properties of peat bogs have yielded the equivalent of an ecological library, legible to anyone with the proper training, telling us about earth's past.

Sphagnum, the Magic Moss

Preserved branches, pollen grains, or corpses owe their relatively undecayed status and common orangey-brown hue to acids in the water called humic acids. Humic acids come from the decay of leaves, twigs, and other dead plant materials; it is what gives soil its dark-brown color. Some decomposition does take place, and in the process acids are released. Acid also leaches from sphagnum moss. Sphagnum moss is clearly the keystone species of a bog—a key ingredient in the chemical cocktail of the bog, and the lattice of its spongy architecture. *Sphagnum* species are common in many types of wetlands, but where the climate is cold and the water is low in nutrients, it takes over and grows so abundantly that it controls water quality, changes water flow, and contours the shape of the land. Like the beaver, sphagnum is an ecosystem engineer, creating large-scale changes in the physical and biological environment (see chap. 3).

Given sphagnum's dominance in a bog landscape, it is strange to learn that the moss is not present at the outset of bog formation, which proceeds much like the wetland-to-upland developmental sequence described in the first chapter. In a true bog, the growing plants live in low-nutrient, acidic water, originating primarily from rainwater and snowmelt. These "real bogs" can only be found in cold, wet climates, in Canada, northern Europe and Russia, parts of the British Isles, and the northern sections of Minnesota, Wisconsin, Michigan, and Maine. Bogs begin in low-lying areas lined with silt, clay, and other water-holding soils. Rainfall, surface water, and groundwater flow in and create a lake. Algae and aquatic plants grow in the lake, and as they die back each winter, dead leaves and stems sink to bottom. Year after year these materials pile up, raising the ground level through the water column. Gradually, the lake fills in, until it is shallow enough for emergent wetland plants such as rushes and sedges to move in. At this stage, if groundwater is still seeping into the basin, the area would be considered a fen (see box 3).

The sedges and rushes of the fen add to the yearly accumulation of dead plant materials (peat) on the bottom of the basin, and the ground eventually becomes more and more elevated over the water table as the peat builds up. When groundwater, with all its nutrients and minerals, can no longer seep all the way up to the elevated peat surface, sphagnum moss can take over. Once established, it works its chemical magic by removing nutrients and adding acids, narcissistically creating the conditions that favor its own growth. The situation snowballs as more sphagnum gives rise to more sphagnum, and a perched bog is born. Bog-like plant communities may also originate in acidic groundwater or low-nutrient surface water, but these are not considered "true bogs" because the plants are affected by more than just precipitation. Other bog communities can form on formerly dry upland areas. Hydrologic changes in the landscape, or persistent humid conditions, cause the land to get wet and soggy, allowing peat to form. Eventually, sphagnum and other bog plants blanket the hillsides, as one can see in the mires and moors of Scotland and Ireland.

Sphagnum species conduct their acidifying alchemy with the help of ordinary rainfall. Rainfall does not carry as many acid-neutralizing nutrients and minerals as groundwater or surface waters do, and as a result it is more acidic. *Sphagnum* cell walls contain humic substances that release acids and gobble up nutrients from the water—particularly, calcium (such as that found in bones, or merely dissolved out of rock) and nitrogen (Crum 1992). Sphagnum hogs all the nutrients and produces highly acidic water that it can tolerate just fine; meanwhile, other species can't survive in these acidic, low-nutrient conditions, so the more sphagnum moss there is, the more there will continue to be as it creates favorable conditions for its own growth.

Unlike rainfall, groundwater and surface water carry lots of key elements from the surrounding soil and the rocks below. If these minerals could reach the wetland, not only could they nourish other plants, they could also neutralize the acids in rain and snowfall, thereby counteracting the effects of the moss. But the physical structure of *Sphagnum* species prevents this from happening. All species of *Sphagnum* contain many large, empty hyaline cells, which absorb a lot of water — as much as twenty-seven times the plant's weight (Crum 1992) — thus explaining its former utility as a diaper for Native American babies. These cells allow sphagnum moss to act as a sponge, keeping its immediate area wet and preventing water from moving. A large growth of it acts as a dam, blocking groundwater flow from below and obstructing mineral-rich surface water from coming in from the sides. As it does this, it grows from the top, leaving the earlier generations of itself behind, dead but undecayed and still holding onto the water and nutrients it absorbed earlier. This builds the elevation of the bog above the level where groundwater or surface runoff can reach the top of the wetland soil. That's important because with sphagnum, and without groundwater or surface water, essential elements like calcium, phosphorus, and nitrogen, are in extremely short supply. And which plant grows well in these conditions? Sphagnum mosses. Thus the sphagnum juggernaut continues unabated.

Small, isolated bogs can form in glacial depressions called kettle holes (although some argue that these are never true bogs because they have groundwater flowing in from the sides), and are found as far south as southern New Hampshire. Bogs can also form along slopes, and on impenetrable soils in foggy, coastal areas, such as in coastal Maine and Ireland. While the final structure of the wetland is not the same on a slope or coast, the overall formation process is somewhat similar. Many bogs formed in this way when the climate became colder and wetter, around 500 to 600 BCE; the *Sphagnum* species actually grew out of the wetlands, creeping up into the forest and croplands. In very cold areas, a raised bog can form. Because of its water-holding capacity and talent for acidification, sphagnum moss grows faster in the middle than at the edge of the bog, forming a dome. The edge of the bog has more air and more minerals from the surrounding upland soils, so decomposition breaks the moss down (Crum 1992). In the middle of the bog, conditions are too cold, wet and acidic for decomposing organisms to do their work — it is the sweet spot for the *Sphagnum* species. There the moss grows happily, leaving vast quantities of undecayed dead material underneath. A few centuries of this yields a large dome or mound where there had only been a basin.

How can sphagnum thrive in conditions that few other plant species can endure? Turns out, these mosses get a lot of help from their friends—billions of them. Recent research has shown that the surfaces of sphagnum (and particularly the hollow hyaline cells that hold the water so well) are colonized by thousands of different types of bacteria. The bacteria help the plants resist dehydration, repair DNA damaged by the toxic chemicals that arise in low-oxygen conditions, protect the moss from disease, acquire and hold onto scarce nutrients such as nitrogen, and even contribute to genetic exchange between mosses (Bragina et al. 2014). The bacteria are passed on from one *Sphagnum* generation to the next on the spores, which are these primitive plants' version of seeds. While these bacteria do exist in the surrounding peat soils and function as helpers for other kinds of plants, they pack themselves into much denser and more diverse colonies on sphagnum mosses than in other microhabitats. A study of just one species, *Sphagnum magellanicum*, found twenty-eight different functional groups of bacteria, with each functional group likely encompassing dozens of individual species (Bragina et al. 2014). Given the "plastic" ability of the bacteria to change their functions in response to environmental change, one can only hope they will be able to assist in sphagnum's survival in the face of climate change.

Beyond their ecological importance, all three hundred or so species of sphagnum moss (McQueen 1990) share a tactile and visual beauty that captivates every bog walker. Sphagnum mosses are soft to the touch, and they make a comfy, if damp, sit-upon. *Sphagnum* species can grow in tangled masses or orderly carpets, pushing up stems topped with stellate or rounded heads, covered with tiny soft branchlets; under a magnifying glass, each branchlet shows overlapping rows of minute leaves, ordered like scales on a pinecone. These primitive mosses seem to borrow their colors from the rest of the plant kingdom, from the sweet green of a spring fern, to the yellow ocher of oak catkins, to scarlet-maple red and pale-orchid pink. Different species thrive in different conditions: the ruby-red *Sphagnum magellanicum* is a denizen of true bogs, while pale-green *Sphagnum subsecundum* will be found at the edges of forested wetlands. Some weak-stemmed species will drape themselves in a pool of standing water, while other more robust species grow atop peat hummocks (McQueen 1990).

Thousands of years of accumulated sphagnum moss imparts a wonderfully squishy sponginess to the bog. To walk on an open bog mat—whether in a true bog or a lake-edge fen—is the kind of joyful experience usually reserved for children in inflatable bouncy houses. A boardwalk resting on the watery bog mat quakes up and down as if it were a trampoline, amusing even the crankiest of curmudgeons.

This is one of the many reasons that a field trip to a bog it is guaranteed to be a good day. In wetland scientist Ingeborg Hegemann's class of engineering students from Lowell, Massachusetts, many students were from underrepresented groups or foreign countries, and few of them had ever visited any kind of wetland. A tall engineering student from Nigeria found the experience particularly captivating. "At first he was fearful, but when he realized he wasn't going to fall in, his face was just beaming—he kept saying, 'I'm walking on water, I'm bouncing!'" she recollects. "He insisted on having his picture taken so he could send it home, to prove that he had walked on water. It was one of those moments you live for as a teacher, when someone is so happy."

On the open bog mat, without the boardwalk, the jostling sensation carries the added thrill of wondering, Will I go through the mat? And then what? The moss in a true bog is usually much denser a few feet down, allowing sufficient traction for an eventual (if undignified) self-extraction, so there isn't too much danger of becoming an archaeological find for the researchers of the third millennium. However, near the edge of the central pool, the moss mat is often only a few feet thick, so it is conceivable that a person or other large animal could go through and become entrapped—or disappear entirely.

Life (and Death) in the Bog

The sphagnum bog is full of cold, acidic water with few nutrients—not a happy place for most creatures. The few plant species besides *Sphagnum* that can grow in these conditions—the bog specialists—show a number of fascinating adaptations. To fight the freezing effects of cold and to conserve scarce nutrients, many bog plants borrow a trick from conifers such as hemlocks, spruces, and firs: they are evergreen. Plants such as the aptly named leatherleaf, bog rosemary (*Andromeda glaucophylla*), bog laurel, and black crowberry (*Empetrum nigrum*) have thick, evergreen leaves that they hold onto year round. Their waxy coverings protect them from damaging cold and wind, and holding onto their leaves through the winter means they don't have to find all the nutrients and energy to grow them all over again in the spring. Come the first warm day with unfrozen ground, they are photosynthesizing while their deciduous neighbors are still building up the energy to grow their first leaves. Plants such as Labrador tea even have furry undersides that may function to keep them warm in winter (Eastman 1995).

It's not just cold in the winter bog—it's wet and frozen. Flooding means no oxygen; no oxygen means no energy; how is a plant to cope? Just like humans craving large hearty meals of pasta in cold weather, bog plants such as large cranberry (*Vaccinium macrocarpon*) use stored starches and other carbohydrates to weather a long, ice-encased winter. Larger plants with older leaves, which contain more stored food, survive the frozen bog better than smaller plants or those with younger leaves (Schlüter and Crawford 2003). The plants of these cold climates also reduce their metabolism as much as possible, to avoid using up their reserves and damaging their internal systems—another adaptation to winter conditions shared with lethargic, couch-inhabiting humans.

An evergreen strategy helps plants survive the bog's low-nutrient situation as well as the cold weather. Even in evergreen plants, leaves (needles, in the case of conifers) do die and drop to the ground, but they don't do it all at once—and right before they do, they suck the nutrients right out of the leaf, back into the living plant. Other plants employ different strategies to corral scarce nutrients. When a plant can't find the food it needs in the soil, air, or water, it is time to go on the hunt.

Carnivorous plants are commonly found in peatlands, particularly pitcher plants and sundews (several species of *Drosera*). The purple pitcher plant is found throughout the southeastern United States and many western states, but it is the only pitcher species in the upper midwestern and northeastern United States. The plant sprouts smooth, leathery leaves that turn a rich red-purple color as they age. The leaves are shaped like a cylinder, with a flap-like hood across the top. The hood contains a "hooker zone," which is just as interesting as it sounds. The underside of the hood is covered with a network of blood-red veins that contain nectaries—small structures that release a sweet-smelling syrup to lure insects. Ants, blowflies, beetles, and even the occasional grasshopper get caught in this natural pit trap (Cresswell 1991). Attracted to the nectar, an insect lands in the hooker zone and may even get a bit tipsy on the nectar; as it tries to crawl out, it encounters a slippery surface spiked with a sharp, down-pointing hairs. Eventually, the insect falls and drowns in the water held in the base of the pitcher.

The dead insects are decomposed by specialized bacteria that live in the water and digestive enzymes released by the plant. The pitcher plant absorbs key nutrients—including nitrogen, which is in short supply in the bog—from decaying insects and bacterial waste products. (Tiny sundews take the same basic approach, using sticky hairs to trap and hold insects, but digests them in place without the help of any bacteria.)

As it turns out, the tiny puddle inside the pitcher harbors a marvelous microscopic community. The bacteria that decompose the hapless insects form the base of this community's food web, and they are eaten by several competing invertebrate predators, such as single-celled protozoans and cup-shaped rotifers. These midlevel consumers are eaten by the apex predators, the lions, tigers, and bears of the pitcher trap: mosquitoes, flies, and midges. Among these are specialized mosquito larvae (which live *only* in pitcher plants and do not bite people) as well the aquatic baby stages of a certain fly species and a midge. All three species compete for the tasty critters in the water, and all of them live in the pitcher without being digested by the bacteria or the plant enzymes. The pitcher plant and its residents are a great example of a mutualistic relationship—where the plant and the creatures living inside it help each other out. In this case, the pitcher plants can survive without their helpers, but the tiny aquatic community inside it has nowhere else to live (Cronk and Fennessy 2001).

The Mystery of the Disappearing Moose

In the center of the bog, near the open sphagnum mat where the pitcher plants grow, many bogs have small ponds. These pools are usually the remnants of the lake-fill process that formed the bog in the first place. The pools are sometimes very deep, and contain "false bottoms"—a murky depth of fluffy peat that looks like it might be solid. But looks are deceiving, as Professor Ron Davis of the University of Maine–Orono describes: "If you tried to stand on it, you would sink all the way down." Working near one of these pools on a cool fall day in central Maine, Ron and his assistant observed the treachery of the bog pools in a place called Caribou Bog. "We were on our hands and knees, quietly counting the plants and recording them on data sheets. We looked up and there was this big bull moose. It seemed as if he hadn't noticed us—he walked right by, five to ten feet away. He was so big and beautiful, we were astounded, and we didn't move, we just watched him walk by. A few days later, we came back to the same place, and we were horrified to find this moose floating dead in the pool." Apparently, the moose had ventured into the pool, probably to eat the carbohydrate-rich rhizomes of the yellow water lily (spatterdock) growing there. "It couldn't get out—it must have been in there for quite some time, treading water, until it finally died. I guess this is kind of a gory story," Ron says. The soft false bottom of the pool had claimed a victim.

But the kicker in this story comes later. It was late fall when the moose met its untimely demise, and when Ron and his team came back the next season, this huge moose was completely gone. "Our conclusion was that the forces of decomposition were quite efficient, and it must have been incorporated into the new growth of the bog." Although the pH of the pool was about 4—which is very acidic, he notes—the bones are very soluble. "The skeleton of the moose would have dissolved really quickly."

Given the notorious ability of bogs to preserve the flesh of vertebrates, humans and otherwise, other answers to the mystery of the disappearing moose need to be considered as well. Frozen into the pool in winter, the moose would have become accessible to scavengers—coyotes are the most common vertebrate predators in bogs, and crows could have helped themselves to many servings of moose meat, too. After becoming part of the local food chain, what was left of the moose could have simply sunk, its bones quickly dissolved, its flesh preserved in the soft and squishy peat.

Moose, and many other large mammals, don't find a lot of food in a bog. Since all the nutrients are sucked into the *Sphagnum*, where it remains locked into the undecayed plant parts for hundreds if not thousands of years, other plants have a hard time growing. Add in the acidity and the cold and the wet, and it is clear why the bog is not a place of abundant, lush plant growth. The bog may *look* plant-packed, but there isn't much growth from year to year. Little plant growth means few animals eating the plants, and those plant eaters are usually the prey for larger animals. There are no large mammal species that spend their whole life cycle in bogs. Generally, moose visit only those bogs that have some aquatic habitat, preferring the smaller peatlands with some forest nearby, particularly when there are willows and birches in the mix (Berg 1992). Caribou (and presumably moose) even take advantage of the unpopularity of open peatlands, using them as refuge to escape their predators—they are less often killed and eaten in bogs than in adjacent forested uplands (McLoughlin, Dunford, and Boutin 2005). Paradoxically, the top predators—mountain lions in some areas—also use the bogs to hide from their own nemesis: humans (Berg 1992).

Lured out to the bogs by their beauty and his fascination with the unusual ecology of bog plants, Ron Davis spent twenty-eight years studying the bogs of Maine, Canada, and Europe. Upon retiring, he wanted to share this experience with more people. Fortunately, the perfect situation existed just outside of his university town of Orono, Maine—the aptly named Orono Bog. Bringing the public into the bog would prove to be a lofty goal indeed: to get to the central bog area, a boardwalk would need to be five thousand feet long and handicapped accessible over the spongy, slightly sloping surface of the bog, all without damaging the bog.

"Too dangerous!" responded one elderly gentleman who attended the Orono planning board meeting to vigorously oppose the boardwalk. Envisioning the peat as a soupy quicksand, this local resident feared that people would fall off the boardwalk and be sucked into oblivion, and then embalmed by the acidic peat. To support his stand, he told the story of his neighbor's horse who ventured to the edge of the bog, got stuck, and died. (Maybe he knew about that moose, too.) Despite this dire warning, the planning board approved the boardwalk. In fact, the Orono Bog Boardwalk idea was met with almost universal enthusiasm. A planning group helped design, fund, and build the boardwalk.

As Ron Davis describes it, many boardwalks are suspended over the surface on pilings, but the twenty-foot-deep peat precluded that design, so they used dock floats to float the boards on the peat. "We built it like they built the east-west continental railroad, where there was a crew coming from each direction and they met in the middle. We even had prisoners from the Charleston Correctional Facility, a minimal security prison, who volunteered to build the eight-by-four-foot-long wooden boardwalk sections, as well as AmeriCorps volunteers from all over the country. We'd put each two-hundred-pound section on a cart, direct it over the constructed sections to the end, and hand it to a group standing in the peat." There are 509 of those eight-foot sections in the boardwalk. "We used serrated bread knives to level the peat surface to float the boardwalk sections. The boardwalk formed a loop, so we had a crew coming from each side of the circle. By late November, the weather was getting worse and worse, and things were freezing up. It was the end of the season, getting pretty close to Thanksgiving. We were going to lose the AmeriCorps volunteers, so we had to get it done. I remember it was in the middle of a snowstorm, cold and windy, when each end of the circle met out in the middle of the expanse of the open bog, and, just like when they constructed the first railways, we drove in the golden spike to finish it!" Even twelve years later, Ron's voice resonates with the pride of that accomplishment.

The Orono Bog Boardwalk is a great place to experience a true bog, as many people quickly discovered: there were twelve thousand visits the first season, and there are now more than thirty thousand per year. In fact, it is so popular that it was completely rebuilt in 2014–15, using rot-resistant recycled plastic lumber, and once again using an all-volunteer construction crew.

Just Add Groundwater to Make a Fen

Heading out to one of the most famous wetlands in the Midwest, Doug Wilcox mused on what he had been told about the area from a group of concerned citizens. Members of the Save the Dunes Council were worried that the water of their beloved Cowles Bog was becoming polluted. Located on the shoreline of Lake Michigan in the Indiana Dunes National Lakeshore, Cowles Bog became well known in 1899, when Henry Cowles, considered one of the fathers of the science of plant ecology, published a study of the area. The wetland is nestled in between the sand dunes formed thousands of years ago when Lake Michigan spread out over a larger area, leaving waves of fine sand behind as its water levels slowly receded. The beauty and unusual ecology of the area brought in many nature lovers in the early 1900s. Unfortunately, the easy access to a major body of water and the proximity of several urban areas also made the location attractive to industrialists, and the fight between development and preservation began.

In 1980, at the time Doug first studied the bog, local environmental activists noted that the bog was not nearly as acidic as bogs are supposed to be, because, they claimed, contaminants were drifting into the bog from nearby industry—particularly the Bethlehem Steel mill just to the west of the wetland. Doug, a wetland scientist with the US Fish and Wildlife Service, and his colleagues suspected otherwise: they knew that a stand of northern white cedar grew on the site, and cedars don't grow in the wet, acidic conditions of bog peat. The area was probably a fen, Doug thought, which is different from a bog in many important ways. Fen water is less acidic, and often has high mineral content. In addition, a key component of a fen is groundwater input, which may be present in the formation of a true bog, but is never able to reach the top of the bog once *Sphagnum* is established. Doug and his team suspected that there was lots of groundwater input into Cowles Bog, which would explain the high mineral content and the lack of acid that local neighbors feared was caused by pollution. But they needed proof. So they consulted maps, gathered equipment, and headed out to Cowles Bog.

"Getting in and out of there is hell," Doug says. "You have to slog your way through cattails, there's poison sumac along the way, there are holes you sink into up to your waist, it's easy to get stuck." To determine if there was groundwater inflow, Doug and his team had to install a special pipe called a piezometer into the wetland, pounding it in vertically like a well pipe. "There was a small mounded area in the bog, and one part of it had northern white cedar," he explains, noting that this tree species has an affinity for groundwater-soaked areas. "We put in the first piezometer right in the northern white cedar stand, and we slammed it in inch by inch. And once we got it down about ten feet, water was flowing out the top"—evidence that groundwater pressure was pushing up into the wetland, like an artesian well. "When we saw that, we just started cheering—it was confirmation of what we suspected. It was a fen, not a bog."

A nuclear power plant had been proposed in the immediate vicinity of Cowles Bog—er, Fen—so evidence that steady groundwater inflow was a critical component of the site got Doug thinking. Having grown up trapping muskrats to make money during his teenage years, he had developed strong natural instincts about wetlands. With a degree in aquatic science and lots of on-the-job training in hydrology, he became one of the first "eco-hydrologists," assembling information about water sources, water levels, water chemistry, and species distributions to understand the natural dynamics that sustain wetland ecosystems. This background is what helped him realize that the more imminent danger to the groundwater-dependent

Cowles Bog was potential dewatering resulting from the planned construction of the nuclear power plant nearby. As it turned out, these concerns led to a delay in the construction of the power plant, and eventually the project was abandoned for other reasons. Years later, activities related to adjacent industry were identified as one of many threats to this unique wetland, especially water-level changes that allowed cattails to invade sedge and grass meadows of the fen. Today, the National Park Service is actively working to restore natural processes to the fen.

Fen . . . it's not a familiar term to most people. But Fenway Park—ever heard of that? The famous baseball field in Boston? It's not the only landmark in Boston metropolitan area named for the extensive wetlands that once covered the land west of Beacon Hill—there's the Back Bay Fens, an urban park and bike path designed by the famous landscape architect Frederick Law Olmsted (see chap. 8), and the West Fens, a shopping area along Boylston Street, to name a few. Most of Boston's Back Bay is made of fens, marshes and mudflats that once served as overflow for the Charles River and groundwater seepage areas, but were filled in long ago to make streets, apartment blocks and parking lots.

For Richer, for Poorer: All Types of Fens

Fens are a type of wetland that, like bogs, have deep organic peat and are constantly wet. Unlike true bogs, they do not need to be located in cold, northern climates, they are not always acidic, and they receive a lot of groundwater flow and possibly streamflow as well. Because groundwater is not the same everywhere, fens are not the same everywhere. Where the groundwater is naturally acidic and nutrient poor, the wetland fed by that water is called a poor fen. Poor fens are hard to distinguish from bogs because the plants that live there are bog-type species that can tolerate those bog-like conditions. Canoeing around the perimeter of a lake or exploring a small headwater stream, you may encounter a floating mat of sphagnum moss, leatherleaf, wiregrass sedge, and even some lovely pink orchids called rose pogonias (*Pogonia ophioglossoides*). Getting out of the canoe and onto the floating mat is an adventure not to be undertaken lightly, so if you want to explore the wetland, you'd best get to it from the upland. In doing so, you'd probably find indicators of groundwater influence at the landward edge, such as tussock sedges and sweet gale; but in the middle of the spongy moss mat, you might even find pitcher plants and white wispy tufts of cotton grass catching the late-summer light. It looks like a bog, it feels like a bog—but because its water sources include more than just snow and rain, it is a poor fen.

Wildly illustrating the art of the oxymoron, Diamond Bog in Rhode Island is an example of a poor fen. Despite its name, it is neither nutrient rich nor a bog. Officially characterized as an acidic fen, it is carpeted with sphagnum and colonized by a multitude of sedges: slender sedge (*Carex lasiocarpa*), bog sedge (*Carex exilis*), and beaked sedge (*Carex rostrata*), to name a few. Sundews proliferate on open areas of *Sphagnum*, and shrubby areas are packed with leatherleaf, sweet gale, and highbush blueberry. Crossing through a deep moat onto the floating mat section of the fen, Frank Golet, professor emeritus from the University of Rhode Island, would warn his students: "Try to stay apart, and try to stay where the shrubs are tall, where the roots are firm." Despite any hazards, the trampoline effects of the floating mat worked its charm: "By the end of the trip they had grins from one ear to the other."

What's in a name, after all? The important difference is that a poor fen, since it has groundwater and potentially some surface water input, will play a different ecological role in the landscape than a true bog. For example, a poor fen at the edge of a stream will provide some habitat for young fish and amphibians to hide from predators such as mink and otters. "I can remember several foggy mornings where you could hear the otters swimming up towards us, and even though they couldn't see us, they could probably smell us, and we'd hear a snort and they'd turn around," Frank says. A poor fen will have greater plant diversity than a bog because there are more minerals and nutrients in the water; more insects will swarm onto the abundance of plants; more fish, frogs, salamanders, and rodents will feed on the insects, and so on up the food chain. It may be a poor fen, but its food web is richer than a true bog's. In addition, the groundwater flowing into fens can react chemically with potential pollutants such as phosphorus that come in from upstream sources. Nitrogen, another problem for downstream waters, can also be removed in the low-oxygen, highly organic wet soil of fens (see chap. 2). In addition, fens act as outlets for natural groundwater flows, allowing springs to flow through, leaving temperatures relatively unchanged. This is important because the temperature of the groundwater coming out of fens is warmer in winter and cooler in summer, which is important for downstream fisheries. All these factors explain why fens are important for protecting the water quality and habitat value of downstream waters (Bedford and Godwin 2003).

The situation gets much more interesting in fens that are fed by groundwater and surface water that contain an abundance of minerals. Sometimes called "geographically isolated," many of these fens have no connection to surface water from rivers or lakes uphill but do take in overland flow. Considered "medium" or "rich" fens depending on the amount of calcium, magnesium, nitrogen, and other nutrients in the water, these peatlands harbor a different variety of flora and fauna than the poor fens and true bogs.

Here, the key elements in the water are those that dissolve off of limestone, dolomite, and other mineral-enriched geologic formations — notably, calcium, magnesium, and bicarbonate, which is an ion made of carbon, oxygen, and hydrogen. In the northeastern states, these kinds of rocks are relatively unusual, and the "rich" fens are not as heavy in calcium and magnesium as those in parts of the upper Midwest and far western states, where a few cups of water from a fen contains, on average, almost as much calcium as an antacid tablet (Amon et al. 2002). And just like an antacid tablet, the water in these wetlands neutralizes acid, leading to much more plant-friendly living conditions than in the acidic, low-nutrient bogs.

Fens that are connected to streams or lakes or other surface water bodies are generally less mineral-enriched than those that are watered only by groundwater (Godwin et al. 2002). Many of the groundwater-fed fens form where flows of groundwater are focused on a few small areas; this can happen on a slope or in a low spot, where there is essentially an underground "channel" of permeable sand or gravel that allows rapid water flow out of the ground. A spring or seepage area results, which keeps the area constantly wet. In wet conditions, decomposition is slow and organic peat builds up. If the groundwater seepage is strong enough, a small mound will form from the peat, like the one at Cowles Bog. In some cases, there is so much groundwater flowing up that it's like an artesian spring.

It is odd to imagine a hill in the middle of a wetland, or a wetland on a valley slope, but the hidden geologic formations make it all possible. Only with the right layering of the correct kind of rocks, in the appropriate topography and slope, will a rich fen form. For example, in Whitewater, Wisconsin, the hills of the kettle moraine nearby are made of porous sand, gravel, and cobbles left by the glacier. These coarse glacial materials allow copious amounts of rainfall to flow into the aquifer, where it dissolves calcium and magnesium from the dolomite bedrock below. All that mineral-enriched water flows underground and discharges along the edge of the elongated lobe of the moraine. As Quentin Carpenter of the University of Wisconsin, who has studied the fens of the area for over thirty years, explains: "There are springs all along the base of the kettle moraine — they are discharging out the sides with groundwater, weeping, seeping into little rivulets everywhere." Standing on the edge of Whitewater Creek, he describes the vista: "You look out and see several mounds rising out of a plain of classic sedge meadow, like a series of islands rising out of an ocean. Each small hill is about eight to ten feet high, covered with low-growing vegetation and topped with a pool where the groundwater is discharging in specific spots." The mounds form wherever concentrated inflows of groundwater keep the immediate area so wet that decomposition slows and peat builds up.

Because these settings are so unusual, rich fens often harbor rare plant species. One study found that as many as 30% of the plants found in Iowa fens were rare species (Nekola 1994), and there can be as many as five hundred different species in one area. For Jim Amon, a fen expert from Wright State University, this impressive plant diversity is what drew him to study fens in the first place. "The diversity is just incredible. If I stand in one place in the middle of the fen and I stretch out my arms and pivot on one foot, I can touch twenty-five different species in that small area." In a thousand-foot-square area of fen — a space roughly equivalent to half a tennis court — there could be as many as eighty different plant species.

"Fens are really like coral reefs in so many ways, because they are so diverse and they are what I call a 'continuous culture' system," Jim says. As a result of the constant groundwater flowing through, "they are always being flushed. Any waste products that are produced are washed away, and minerals and nutrients are constantly being brought in to replace them. It's like an intravenous feeding system." This flow-through does not occur in bogs, nor does the high plant diversity. "In bogs, water gets in the bowl and stays there," Jim notes, "while fens are fed by flowing groundwater." And, he adds, fens "are really beautiful places."

Rich fens might include buckbean (*Menyanthes trifoliata*), Kalm's lobelia

(*Lobelia kalmii*), or grass-of-Parnassus (*Parnassia glauca*) (Godwin et al. 2002). Some of the rarest fen plants are the fen beakrush (*Rhynchospora capillacea*), hairy valerian (*Valeriana edulis* var. *ciliata*), and the small white lady's slipper (*Cypripedium candidum*). Many of these rare plants are not strong competitors and have evolved mechanisms to survive in consistently wet, calcium-rich conditions that other plants don't tolerate. But if the area dries out because of groundwater diversions or other impacts, then more-common plants that are better competitors take over. The spreading globe-flower (*Trollius laxus*), for example, is found in shady parts of fens where the water level is near the surface, but it would grow just as well in drier, sunnier areas (Scanga 2011). However, in drier, less stressful conditions, the globe-flower is easily outcompeted by taller native plants, as well as by invasive species. In many cases, rare plants are rare because they can survive in unusual settings that other species cannot handle.

Other problems can arise in these rich fens if the water quality is degraded. Even though the fens are high in minerals (mostly calcium, and magnesium) they are low in the nutrients most plants like (nitrogen, phosphorus, and potassium). Invasive plants, such as reed canary grass (*Phalaris arundinacea*), common reed (*Phragmites australis*), hybrid cattails (*Typha* × *glauca*), and others, grow tall, fast, and furious when nutrients come into the area from lawn fertilizers or farm chemicals, or when peat deposits decay and releases these critical nutrients. However, if the water stays clean, then these invasives can be starved right out.

This is illustrated nicely by two long-term observations from Quentin Carpenter. "In 1988, which was one of the driest years in Wisconsin history, there was a cow pasture next to the fen," and a determined rogue steer broke through the electric fence and "found his way to the middle of the highest mound" in Clover Valley Fen. "I chased him off, and then found this huge cow pie, just laid down, right on top of the mound. It was too big to carry out, so I figured I'd leave it there. Next year, I noticed the cow pie looked very, very green—it was reed canary grass." Tiny, tiny sprouts of the invasive plant, brought to the spot in the manure—the cow had been munching on this invasive plant elsewhere. "I left it, figuring that not much would happen. By 1992, from that one cow pie, the reed canary grass had spread about three feet in each direction, right on the top of the mound"—where all the low-growing vegetation and rare plants are found. "I checked it every few years, and every time it was still there, but it never got any bigger. In 2013, twenty-five years later, the reed canary grass is still there, but it is not very tall, and it is down to less than one square meter. That one input, that one burst of fertilizer in its rawest form, was enough to get it going, and once there, it is very, very persistent, but it didn't keep expanding and take over. But if the cows got in there for a long time or if fertilizer contaminated the aquifer—it wouldn't take much of an increase in fertility to tip it in the direction that favors the invasive species," Quentin concludes.

Not far away, in Bluff Creek fen, a 1955 dredging operation started a similar vegetation altering process. Drawing on the recollections of botanist Galen Smith as well as his own observations, Quentin describes the scenario: Removal of the organic peat to try to straighten the creek and drain the wetland left a long line of spoil piles between the creek and the upland. No longer wet all the time, these peat piles began to rot and release the nutrients that had been locked in the undecayed plant materials. This allowed a clone of cattails, connected by long airy roots called rhizomes, to move in, creating a thick, dense stand on the spoils. They had advanced about sixty feet into the fen. "The interesting pattern was that the thickest cattails were in the middle. Where they had been for twenty years they were thin and weak, in the middle they were dense, and then on the advancing end they were thin again. Every few years you could see the densest part move up the spoil pile, kind of like a moving wave," until they reached the upland.

As they moved into the new territory the cattails did well, and grew fat and happy. But as they used up available nutrients, they had to move on, and eventually they died back. "They invaded and made mischief, but did not take over completely," Quentin says. "Cattails have a lot of aerenchyma, and they can pump extra oxygen into the soil to free up some nutrients" by increasing decomposition. "But there weren't that many nutrients to start with, so that trick didn't work anymore." They had to keep moving up until they ran out of wetland area. Without the continual inputs of nutrients, the invasives can't take over. "Groundwater there is steady, steady, steady, it is hard water, very high in calcium and magnesium and low in any nutrients. All the land around it is conservation land, so it doesn't have much pollution," he explains. The groundwater probably flushed out much of the nutrients decaying off the pile.

Because rich fens are so diverse and uncommon, many people have worked very hard to protect them. An example comes from Jim Amon's Beaver Creek Wetlands Association, just east of Dayton, Ohio, where a series of fens wind along a narrow stream corridor, sandwiched between housing developments and farm fields. Through land acquisition and by working with developers, this organization has protected almost all the area along the creek. "There are nine access points, and we have installed educational signs at each place," Jim says. "And each of the nine areas are unique — you can go to each one and see something very different. The main place in this ten-mile corridor is a shallow-mound fen; the peat is about eight feet thick. In some areas we have these mound fens, and in others we have fens at the toe of the slope, which form a mound but then grade down into a marshy area where in the deeper water you get reedy growth. Then, on the tops of the hills, we have shortgrass prairie. There are also some forested fens," he continues. "They are gorgeous. Walking into the forested areas of the fen, it's mostly cottonwoods and green ash now, and there are some silver maple coming in. The area is changing — emerald ash borers, an invasive insect, are killing the green ash; beavers are killing the cottonwoods." As a result, there is an open canopy with more light. "The sedges are moving into these areas, and a sedge meadow is establishing — Gray's sedge [*Carex grayi*] with huge spiky seed heads. It is kind of cool. In one short walk you can see fifteen species of sedges."

Gorgeous, yes, but to the uninitiated, the fen can be . . . well, treacherous. Jim tells about a particularly memorable mishap: "Out in the fen, we have such strong water flow in some areas that it forms what I call 'quick-muck.' In one second you can sink down to your waist. One of the things I tell people: you don't want to go out there alone. One very hot day, I had gone out to an area with a lot of water discharge. I was getting water to water the fen plants I had grown from seed in the greenhouse. So I'm out there with my five-gallon carboy, and I see this bump in the peat. It turned out to be a geology student who hadn't heeded my warnings — he had been stuck for over an hour. By then, he was hot, scared, and practically incoherent, basically a basket case. He had walked in an area with no plants on the surface," he says. Any experienced wetland wader knows to tread *on* the plants and their roots; the enticingly open areas in between the plants are basically watery mud disguised as firm ground. So how did Jim get him out? "Basically, I kind of talked him out, got him to lean over and work himself out until he could reach me; then

I grabbed his hand and pulled him out." It is probably a safe bet that that young geologist did not pursue a wetland-focused career path.

Although Ron Davis's bog in Orono, Maine, and Jim Amon's fen in Alpha, Ohio, are a long way from each other, the careers of these two wetland scholars took similar trajectories. After decades spent enthralled by the crazy connectedness of water, chemistry, soil, plants, and wildlife in the wetlands, they both dedicated their retirement years to efforts to protect and share these prizes with others. Night after night in meetings to determine goals, raise funds, obtain permits; weekend after weekend eradicating invasive plants, erecting educational signs, building boardwalks.

Their devotion is born of a type of love not much discussed in modern society—the love of a place, of the experience it gives and the nature it harbors. E. O. Wilson (1984) calls it biophilia: the evolutionarily hard-wired tendency of humans to connect emotionally to living creatures. But it extends beyond individual species, to the entire landscape. We who love wetlands feel it as a pull as magnetic and strong as that of the bond between a parent and child: We must go to this wet place, see the impossibly blue sky of an early spring day glaring between the spotted gray twigs of wintergreen, reflected in the dark water puddled at our feet. We need to feel the bounce of the thawing peat beneath our boots, and watch the chickadees find spider eggs in the sedges just poking through the rainbow of sphagnum. It satisfies a deep need within us and leaves us sighing in relief and gratitude, as it inspires in us the drive to protect these places for future generations.

CHAPTER 5

Wooded Wetlands: Basin Castles and Big-River Swamps

> When I would recreate myself, I seek the darkest wood, the thickest
> and most interminable, and, to the citizen, most dismal swamp, I enter
> a swamp as a sacred place,–a *sanctum sanctorum*. There is the strength,
> the marrow of Nature.
> –HENRY DAVID THOREAU, *Walking*

Attempting to extract her foot from the muck, a student lands on her rear
with a whump as she loses her balance. "Isn't there an easier way to get to
the spruce swamp than through these hellish shrubs?" she gripes. But there
is no access any less strenuous: the lovely forested wetland we seek to gain
is surrounded, like a castle fortress, by a moat of scrub-shrub. It is literally
a moat, where water running and seeping from adjacent slopes collects
and deepens, allowing various shrubs, with their greater tolerance for the
low-oxygen situation of the deep water, to win out over the trees. The gray
branches of highbush blueberry and winterberry create an impenetrable
thicket at shoulder height, leading the inexperienced swamp trompers to
choose a seemingly easier pathway, free of branchy tangles, only to discover
that these spots are shrubless because they are too wet, sucking boots right
into the treacherous muck.

Scandinavians—who know more about forested wetlands and northern bogs than most—call places like this a lagg, usually referring to the transitional zone between forested bogs and the upland landscape. Hearing this, at least one student always feels compelled to make sad jests about lagging behind, to add to the "bogged down" and "mired in" jokes that abound on these hikes, which for them become endurance tests. Many forested wetlands have similar zones—an edge created by different levels of water, or sometimes by a combination of water and light favoring the bountiful growth of shrubs over taller, slower-growing trees. As we walk, the wet moat becomes shallower, then disappears, the shrubs giving way to a dark forest of skinny spruce and balsam fir trees with open, easily navigable territory underneath.

Our destination is, in fact, not a swamp, but a lovely wooded wetland. When standing knee-deep in water, looking up at a bright green canopy of red maple or green ash, tupelo, or cypress, wondering what manner and number of leeches are regarding your calves as a dining commons, swamp seems like exactly the right word. Squishing lightly across undulating mounds of lime-green sphagnum moss topped with the emerald shamrocks of goldthread plants, under a cathedral of black spruce, conjures a medieval world of pixies and sprites—a wholly different experience than the swamp, worthy of the more lofty-sounding title of forested wetland. Still again, detangling yourself from the shrub thicket surrounding either of these wooded wetlands elicits a completely different set of images, as well as a variety of epithets not at all suitable for the ears of young children.

And yet all of these are considered "wooded wetlands"—ecosystems that do not show up on the public's radar screen when they are thinking about wetlands. Researcher Andy Cole of Pennsylvania State University notes that most people think that an area has to be covered in water to be considered a wetland. "They look for cattails and red-winged blackbirds, and if there are none, they don't think it is a wetland. Forested wetlands are sometimes pretty dry in midsummer, when people are out looking around, so they get overlooked entirely," he comments. "My biggest message to the general public is that a wetland does not always have standing water on it." (See chap. 6.)

Wooded wetlands will form in low spots of any size or any place that stays just wet enough to discourage upland species of trees, but not so wet that standing water collects for long periods of time. Most trees can't tolerate too much wetness, although there are some exceptions, such as the big tupelos and cypresses in the southern bayous and swamps. In most of the rest of the United States, however, wooded wetlands can be roughly divided into three major categories. On the wet end of the spectrum, where water is too deep for most trees but too dry for cattails and their ilk, shrub thickets will flourish (table 2). Then there are the basin swamps, which are found in wet areas adjacent to the smallest headwater streams high up in the watershed, in wet seepy spots along slopes, and in large flat low areas where water doesn't drain out. All the way down at the bottom of the valley, alongside the big rivers, floodplain forests and bottomland hardwood forests of silver maple, green ash, cottonwoods, elms may flourish.

Shrub Thickets

Bushwhacking and the Gift of the Spider

At the edges of the forested wetlands and in the transition zone surrounding many marshes, a woody tangle of dense shrublands grow. A rich lexicon describes these areas: *scrub-shrub, ganderbrush, carr, thicket, bosque, lagg.* With water only ankle-deep or less for most of year, these areas contain an impressive variety of shrubs, each of which tells a different story about the soils, water, and history of the place. The names roll off the tongue far more gracefully than the feet travel through the actual shrubs. In New Jersey the name for these astonishingly dense low-shrub zones is *pushcover,* because the hunting dogs have to push it aside to see what's hiding behind (Lopez and Gwartney 2006).

Many of these shrub-dominated areas will develop where groundwater is seeping out of the aquifer—perhaps in merely damp conditions along a slope, on the edges of forested swamps, or on flats in a stream valley bottom. Thickets of speckled alder (*Alnus incana* subsp. *rugosa*) flourish in streams and along lake edges where the water is nutrient-enriched, often accompanied by red osier dogwood (*Cornus sericea*), silky dogwood (*Cornus amomum*), poison sumac (*Toxicodendron vernix*), and willows (*Salix* spp.). In these groundwater-fed areas, a thick, orangey goop sometimes forms in the water. Alarming as this looks, it is just iron-rich groundwater reacting with

Table 2. Types of wooded wetlands

Wetland type	Location	Water depth	Most common plant species (varies by region and local conditions)	Typical wildlife
Shrub thicket	In or along the edges of small streams, lakes, or low spots	0-2 feet (0-0.6 meters)	Water willow, buttonbush, speckled alder, silky dogwood, willows, leatherleaf, highbush blueberry, spicebush, arrowwood, wild raisin	Woodcock, ruffed grouse, spotted turtle, four-toed salamander
Basin swamp	Wet areas near streams or lakes; low spots	1 foot deep (0.3 meters) to 1 foot belowground	Trees: red maple, Atlantic white cedar, northern white cedar, black gum, black spruce, tamarack, hemlock Shrubs: blueberry, meadowsweet, spicebush, sweet pepperbush, winterberry	Wood duck, hooded merganser, northern waterthrush, gray tree frog, dusky salamander, masked shrew, water shrew, deer, moose, bobcat, black bear
Floodplain forest/bottomland hardwood forest	Floodplains of large rivers	3+ feet (0.9+ meters) deep in spring; 3+ feet belowground summer-fall	Silver maple, willows, cottonwood, tupelo, cypress, green ash, black ash, American elm	Warblers (cerulean, prothonotary, American redstart), kingfisher, goose, evening bat, Indiana bat, crayfish, heron, mink, muskrat, beaver, otter

oxygen; and, in the presence of certain bacteria, a slimy particulate of iron hydroxides will form. It is a great indicator of groundwater presence, so this type of shrubby wetland would also be considered a fen (see chap. 4).

In places where the water is deeper and poorer in nutrients, a shrub-carr community forms. Here, the most common low shrubs may be the ubiquitous leatherleaf (*Chamaedaphne calyculata*), with its neat rows of evergreen leaves, shiny forest green on top, and rusty-red leather on the bottom, as well as meadowsweet (*Spiraea alba*). Leatherleaf and sweet gale (*Myrica gale*) often team up to form a deep, spongy mat at the edge of lakes and rivers. Characteristic tall shrubs include highbush blueberry (*Vaccinium corymbosum*), inkberry (*Ilex glabra*), and maleberry (*Lyonia ligustrina*, which must win a prize for the most melodious-sounding scientific name). Butterflies and bees flock to the scents and nectar of sweet pepperbush (*Clethra alnifolia*), spicebush (*Lindera benzoin*), bayberry (*Myrica pennsylvanica*), ninebark (*Physocarpus opulifolius*), and swamp azalea (*Rhododendron viscosum*). The silky gray catkins of pussy willows (*Salix discolor*), so well-loved by children, also provide nectar in spring for native bees, and mourning cloak butterfly larvae feast on its leaves.

Where farm fields or recently logged areas are left to grow back, a complex of many shrubs may develop, including elderberry (*Sambucus canadensis*), red chokeberry (*Aronia arbutifolia*), highbush cranberry (*Viburnum trilobum*), and arrowwood (*Viburnum dentatum*), all of which produce tasty berries happily consumed by cedar waxwings, robins, ruffed grouse, and many other birds, as well as small mammals. Many of these shrubs — particularly meadowsweet and winterberry (*Ilex verticillata*) — will grow on the edges of open water, where their submerged stems become critical habitat for nesting four-toed salamanders (*Hemidactylium scutatum*) as well as foraging spotted turtles (*Clemmys guttata*). Water snakes (*Nerodia sipedon*) and ribbon snakes (*Thamnophis sauritus*) may be found wound into the lower branches, basking in a warm Indian-summer sun.

A particularly striking type of shrub zone often develops along the shores of some lakes and ponds, where shrubs like water willow (*Decodon verticillatus*) and buttonbush (*Cephalanthus occidentalis*) thrive in deep water. Buttonbush, so named for its spherical seed heads, can form a dense ring around a shallow basin. Studying one of these buttonbush swamps in the 1980s, wetland scientist Ingeborg Hegemann had an experience she never forgot. "One of my earliest assignments was to evaluate a wetland in Attleboro, Massachusetts, that was proposed to be filled for a regional mall — the project was extremely controversial." As she arrived at the wetland site, the stress permeated her day: it was a sweet little pond surrounded by a thicket, and she could not imagine destroying it for a shopping center.

Nevertheless, she had a job to do, so Ingeborg gathered up her tape measure and clipboard and set to work. As she walked a straight line into the wetland, the ground gradually sloping and the water deepening, she carefully identified and tallied the plants she found. Immersed in her task and her place, she spied something floating out of the corner of her eye. At first it looked like a large leaf. Then it floated closer, and to her horror she realized it was a spider—and not just an ordinary delicate little marsh spider. "It was the biggest spider I have ever seen. It was hairy and gray and it was as big as my hand, fully stretched out. My immediate reaction was that I was just going to faint." Frozen with fear and feeling dizzy, she fought the urge to pass out. "I realized that if I fainted I would fall *in* the water and the spider would be on top of me. I had to talk myself off the ledge and calm down.

"So . . . I just stood still, and I breathed. And as I did so, I realized what an amazing creature this huge spider was—beautiful sleek lines, floating on the water, creating little water-dents around its feet. We eyed each other, neither one of us moving. I'm sure the spider was probably as wigged out as I was. I slowly pushed the water so the spider would float away from me. As I calmed down, I took in the scene—the most brilliant green contrasting with the blue. I could hear the highway in the distance, a very faint hum. I was visiting this magical cathedral, and as I left I realized that there were spiders everywhere. The spider made me more aware, made me stop thinking and being so technical—the gift from the spider."

Upon returning home, she consulted several references and found out that her nemesis-turned-muse was a fishing spider, most likely *Dolomedes triton*. A waxy coating and tiny hairs on its legs allow it to walk on water—or, rather, row through the water—as fast as three feet per second, quick enough to jump on any unwary small fish or tadpole, or a drowning wetland scientist. Often found on the sides of boat docks as well as in dense wetland vegetation, these clever hunters lure in prey as much as five times their body weight by tapping on the water surface with their legs, imitating a small insect. Striking quickly, they bite their victims and immediately inject venom that paralyzes the prey and dissolves the body parts into a sushi smoothie, easy for the spider to suck right down. Given this gruesome talent, Ingeborg's arachnophobia seems like a pretty rational response; but these spiders pose no threat to humans, and, like any predator, they play an important role in the food web. While an ecologist like Ingeborg appreciates these roles intellectually, sometimes it takes a close—and fearful—encounter to develop a richer understanding of "the strength, the marrow of Nature," as Thoreau notes in the chapter epigraph.

Basin Swamps: Wet Forests of Headwaters, Slopes, and Bottomlands

Sanctuary, I: Cool, Fire, Heat

A hot day in July. Heat shimmered off the new pavement; exiting the air-conditioned truck seemed like calculated suicide. Before he had even grabbed his field gear from the back of the truck, ecologist Rick Van de Poll could feel the sweat beading up on his forehead. Stepping off the road bank, leaving the rushing cars and the searing pavement behind him, Rick entered into the wet forest in Rindge, New Hampshire, immediately experiencing a dramatic shift. The temperature dropped, the swamp's cool, moist air enveloped him, and the road noise fell away as birdsong reached his ears. With each step, the fragrance of trees and damp earth replaced the caustic odors of the highway.

Rick's enjoyment of this welcome change was quickly forgotten as he explored the forest glade. "Immediately I knew this place was different. I could tell that this swamp was old, and undisturbed. There were no cut stumps; the trees were large—as much as two feet in diameter, which is pretty big for a red maple growing in wet conditions." Next, he saw a huge rotting log, almost thigh-high, lying on its side. A thick covering of mosses and some small tree saplings growing out of it clearly showed that this downed tree was serving as a "nurse log," providing a sunlit open spot for tree seeds to sprout. He looked around for the typical evidence of agricultural use or timber harvest—ditching, draining, livestock ponds, channelized streams, stumps, stone walls—and found none.

Through Rick's practiced eyes, he could see that the forest had never been touched by agriculture or logging, which is very unusual for southern New England, where 75% of all forested areas were cut down at the height of the farm era in the 1800s. "It dawned on me—this is effectively an old-growth site, a primeval system. There was no evidence of direct impacts. I started to think, if that's the case with this forested swamp, why wouldn't it be the case for many of the wet forests that were too wet, too mucky, or just didn't have enough valuable tree species such as cedar."

Red maple swamps, cedar swamps, and spruce-fir swamps are all forested wetlands that form in low, wet places, but not alongside large rivers. Such low places may be "kettle depressions" formed from blocks of melting ice left by a glacier, or found where groundwater seeps out along the bottom of hillsides, along small stream corridors, or simply in the lowest spot on the landscape. In each situation, most of the water comes from rainfall, with varying amounts of groundwater flow and input from smaller streams and rills. And once in, most of the water doesn't flow out. Instead, the water either evaporates, is sucked up by plant roots, or just sits there, stagnant, leading to a buildup of organic matter—thick layers of black, mucky soil.

In basin areas that have little groundwater input and little streamflow, the plants that grow are those that thrive in nutrient-poor conditions—particularly evergreen conifers such as red spruce (*Picea rubens*) or black spruce (*Picea mariana*), eastern hemlock (*Tsuga canadensis*), and balsam fir (*Abies balsamea*). Such places are particularly common in the northern, cooler climes, and are often underlain by deep organic soils. Add in a little more streamflow and groundwater, and the more nutrient-loving species such as red maple (*Acer rubrum*), black ash (*Fraxinus nigra*), green ash (*Fraxinus pennsylvanica*), northern white cedar (*Chamaecyparis thyoides*) and tamarack (*Larix laricina*) will thrive. These "enriched" wetland areas may also be home for many rare species such as showy lady's slipper (*Cypripedium reginae*), especially when groundwater flow brings in key elements such as calcium.

The most commonly encountered tree in most depressional wetlands is the red maple. Taking over in many parts of the Northeast after the valuable Atlantic white cedars were cut and removed (Rheinhardt 2007), the red maple tree is now common both in upland and lowland woods. In central Pennsylvania and Virginia, for example, it is now found in every type of forest, whereas it used to make up less than 6% of trees in these states. The red maple tree turns out to be a "super-generalist" (Abrams 1998); it can survive in dry, damp, or very wet soils. In fact, blights and insect outbreaks that have devastated chestnuts, elms, hemlocks, oaks, and ashes have left the red maple unscathed. Red maple can also survive extended flooding longer than other wetland tolerant tree species, such as American elm, river birch, and sycamore (Jones et al. 1989, in Cronk and Fennessy 2001).

However, as tolerant as they are, red maples cannot survive fire. Unbeknownst to many people, flames started by lightning strikes are an essential component of many ecosystems. Many plant species are adapted to a light burn and will not thrive without it. We don't think of wetlands as capable of burning, but they are, and they do. For example, in northern Michigan, pollen records show that conifer swamps burned about every three thousand years, and many other wetlands burned much more frequently. Atlantic white cedars are just one of the species that thrive after a fire (Kost et al. 2007); in the Great Dismal Swamp in Virginia and North Carolina, which used to be dominated by Atlantic white cedar, various sections burned as often as every twenty-five years or as rarely as every three hundred years (US Department of Agriculture 2005). Since people tend to respond to any fire, whether natural or not, with buckets, hoses, and flame retardants, the lack of fire has contributed to the decline of Atlantic white cedars and the rise of the red maple.

If the swamp has some groundwater inputs, it is not uncommon to encounter skunk cabbage (*Symplocarpus foetidus*) poking purple-and-green-streaked hoods through the gray and brown duff—or even through a thick carpet of snow—along seepy swamp-side slopes and stream bottoms. Skunk cabbage greens up early in the year because it can generate its own heat. Hidden inside the pointed hood, the fleshy flowers are grouped together into a spadix. The spadix, which looks like a small, martian-green football, decorated at regular intervals with four-pointed stars, produces more heat as the thermometer drops, ensuring that its temperature stays as high as 86 degrees Fahrenheit (30 degrees Celsius) (Seymour 2004). This type of thermoregulation (temperature control) is usually reserved for animals. How exactly the cells of the flower produce heat is not clear—they do not convert sugars into energy, but seem to have a complex biochemical pathway involving enzymes deep in the cell structures (Seymour 2004). The heat helps broadcast its trashy signature scent, attracting a certain class of pollinators—flies, honeybees, springtails, thrips, among others—to the warm, garbage-odored chamber inside the hood (Rice 2012).

Before spring is in full swing, rosettes of the long-stalked, white-veined skunk cabbage leaves cover the ground, creating a great green cabbage patch standing twelve inches tall or more. Although the putrid smell of the crushed leaves discourages some animal grazing, at least one important animal is not deterred. Hungry from a long hibernation, black bears will eat the starchy roots and crunchy stalks of the new green leaves in early spring, particularly in years of low acorn crops (McDonald and Fuller 2005), along with shoots of sedges and marsh marigolds and any other newly sprouting plant. Because of its importance in the food web and its sensitivity to disturbance, the presence of skunk cabbage is an excellent indicator of the overall health and ecological integrity of a wetland (Stapanian, Adams, and Gara 2013) — a swamp with skunk cabbage is, in other words, in fine condition.

Back in the red maple swamp in New Hampshire, Rick Van de Poll reflects on other aspects of ecological integrity: intact ecosystem processes, a fancy term for the complex interactions of air, water, soil, plants and animals that take place in these wet forests. Where trees aren't harvested and where sources of water are clean and keep the area wet, organic material accumulates quickly, creating a permanent repository of the carbon that the plants pulled out of the air in photosynthesis. More carbon locked up means less carbon in the air to contribute to climate change. "That old-growth red maple swamp was a turning point for me, in my understanding of why I was working to preserve wetland resources: for these ecosystem processes," such as carbon storage and diverse microhabitats, that aren't found in a younger or more disturbed wetland. "I've been a naturalist since I was four years old, when I drew my first book of leaves. I've been swamp walking since 1985; I've seen a lot. But walking into that red maple swamp, so close to a highway—it was like a sanctuary."

Red maple swamps are quite common, but one variation of this wetland type is rather rare—the black gum (*Nyssa sylvatica*) swamp. Nicknamed the "bung tree" because they were used to make the plug (or bung) for wooden barrels, black gum trees like their feet wet, and are often found in standing water. Other than serving as material for barrel plugs, black gums have been considered nearly worthless as a forest product (and also have little medicinal importance, unlike the valuable sweet gums), which explains why they were never cut down. Tom Howe, ecologist with the Society for the Protection of New Hampshire Forests, describes the day when he found a stand of old-growth black gum trees: "I was dutifully walking the property boundary of a parcel we were going to protect, occasionally crossing waist-deep water, when I came upon the trees. They are typically dispersed among red maples, with lots of shrubs in the understory. Later I brought the landowners in to see them, and they were so thrilled to have these ancient trees on their land—the largest tree was about twenty inches in diameter, which makes it about three hundred or more years old. The really old trees have deeply furrowed bark, long craggy crevices as deep as three inches." In fact, the oldest known hardwood tree in North America is a seven-hundred-year-old black gum tree (Sperduto and Kimball 2011).

Color-Changing Frogs and Untamed Shrews

Large old trees grow and die, and the light gaps that result lead to new growth, creating a forest of many canopy layers and differently aged trees that supports a great number of species. Most basin swamps are full of humps and bumps, which form from moss-covered downed logs, bases of trees, and rocks, as well as from the tussocks of sedges or other plants seeking to grow up and out of the water. Such topography provides islands of drier hummocks between the wet hollows, creating wet and dry niches for all manner of secretive beings. We expect sprites, elves, gnomes, and otherworldly folk in these mossy glades, but if we are really lucky, we find four-toed salamanders. These finger-length brown salamanders with shiny white bellies lay clusters of eggs on the underside of *Sphagnum* mosses adjacent to pools of water; after six weeks or so, the tadpoles wriggle out of their jelly eggs and drop into the water. Darting pixie-like along a stream's edge, the tiny brown-and-white-feathered northern waterthrush (*Parkesia noveboracensis*) is another common but secretive animal of the forested swamps (O'Connell et al. 2013). These small birds will actually dash in and out of woodland stream rivulets to catch stoneflies and other aquatic treats.

More thrilling than any wee fairy, perhaps, is the gray tree frog (*Hyla versicolor*). Employing sticky-pad feet, they climb into the tree canopy and surprise us with their burring call; although it is a mating call produced only by the males, they seem to particularly like calling as the barometer drops— expect some rain when you hear their staccato trills from the tops of trees! You'll have to rely on your ears rather than your eyes to find them: in only thirty minutes, gray tree frogs can change colors, blending in with light- or dark-gray tree bark or green leaves as needed.

Listening to the tree frogs is another shy creature hiding in the hummocks below: the masked shrew (*Sorex cinereus*). Although these tiny animals do not live exclusively in swamps, they thrive in pockets of damp moss, which are particularly abundant in forested wetlands. Weighing in at only one-eighth of an ounce—less than a teaspoon of sugar—and with a body measuring only two and half inches long, the masked shrew is one of the world's smallest mammals (Whitaker 2004). With its brown-gray fur and long tail, it is often mistaken for a rodent, but its tiny eyes and pointed snout mark it as an insectivore, a carnivorous group of mammals that includes moles and hedgehogs as well as shrews. Because they are so small, the masked shrew, like other shrews, needs to eat every few hours; in fact, they eat about three times their body weight every day. Cagey predators, they are quick to snarf up any bug, caterpillar, worm, slug, snail, or spider that crosses their path.

You are not likely to see masked shrews; they move around mostly at night, and mostly in mazes of tunnels connecting small holes at the bases of trees or hummocks where they build their nests. Baby shrews stay in the nest for about thirty days, emitting high-pitched chirps at their harried parents in hopes of being fed. When disturbed by an intruder, young shrews run away in a "caravan" system, with each youngster burying its snout in the fur of the shrew in front of it as a way of keeping the line together, compensating for their poor eyesight (other shrews form a caravan by biting onto the tail of the one in front!). Like many mammals, they have special touch-sensitive "whiskers" called vibrissae on the sides of their faces and along their feet, which help them navigate in the dark. The masked shrew doesn't like to swim—unlike its cousin, the even smaller water shrew (*Sorex palustris*), which has special adaptations for swimming and diving, and can capture fish, tadpoles, and even crayfish.

Larger animals, such as deer, moose, bobcat, mountain lions, and others also take advantage of the excellent cover of the swamp's dense foliage as well as the rich buffet of rodents, amphibians, birds, bugs, and berries found there. Prior to European colonization, caribou, elk, and other megafauna would have found sustenance and sanctuary in and along these swampy lowlands. And where the animals went, people followed.

Gimme Shelter, Food, and Clothing: A View into the Past

The first discovery was a flake—not a fluke, but a flake—a thin, angular shape made of a lustrous dark red rock called chert. For over a week, Bob Goodby and his archaeological team had been screening shovels of sandy soil from a high terrace adjacent to Tenant Swamp, a forested wetland in Keene, New Hampshire. The tiny piece of stone looked like a flake, a piece of debris from stone tool making. "It had been so long since I had seen one, I wasn't sure—I showed it to my colleagues and they said, 'Duh, it's definitely a flake!' I went back, threw a few more shovelfuls of soil in the screen and there were twenty-five of them. One of the crew said these look just like what she had been finding at Paleo-Indian sites up north. We had a real epiphany at that point: it turned out to be the oldest Paleo-Indian site in the state of New Hampshire."

The site was perched on the steep edge of a large forested wetland along the Ashuelot River, a tributary of the Connecticut River. It was mere months away from destruction for the construction of the local middle school. More testing yielded more artifacts, making the site eligible for registry on the National Register of Historic Places—to the chagrin of the local school board, whose project could not be completed until the site could be studied and all the artifacts retrieved.

Given the landscape setting, Bob, a professor of anthropology at Franklin Pierce University in Rindge, New Hampshire, was not entirely surprised to find evidence of Native American presence, most likely the ancestors of the modern Abenaki tribe. "We often find Native American artifacts on the edge of wetlands. There is a large variety and density of resources in these northeastern wetlands—not just foods but medicinal plants, cattail fluff used in diapers, and so on," he says. Native Americans cut strips of black ash, a common wetland tree, to weave into baskets. Beaver pelts were tanned for clothing. Willows provided painkillers—the source of the active ingredient in aspirin even today. A high, sandy, well-drained terrace near a wetland is a promising location for archaeological discoveries.

"Twelve thousand years ago, the Ashuelot River was lapping at the bottom of the terrace, just a stone's throw away. To the south, there were probably marshlands, and lots of spruce, pine, birch, and poplar forest on either side," Bob says. For these native people, it would have provided good visibility and easy access to the river for travel by canoe.

A full-scale archaeological dig was eventually launched, documenting four tent sites, each with a central hearth and featuring over two hundred stone tools over the whole site—scrapers, gravers, wedges—as well as burnt animal bone, probably caribou, and lots of stone flakes from tool making. Radiocarbon dating showed that the artifacts were abandoned 12,600 years ago, during the period called the Younger Dryas—a bitterly cold time (Goodby et al. 2014). Bob explains the pattern of the materials that were found: "All of the artifacts were found inside the house floors— none of them outside. At other sites, you find butchering areas, tool-making areas. Here everything was indoors, and there were almost no hunting implements; I think this was a wintertime site. Winters during the Younger Dryas had to have been brutal. These people are hunkering down; they are eating stockpiled foods, making clothes, repairing tools, getting ready for the spring." Once spring came, this small group—possibly all women—would have gathered up their belongings and moved to another area, most likely heading downriver to catch the upstream movement of spring migrating fish such as shad and alewives.

Today, what is left of the small gathering area is gone—leveled for the middle school's fine oval track. But the wetland is still there, full of black spruce, red maple, highbush blueberry, and lovely tussocks of fine-leaved three-seeded sedges (*Carex trisperma*) and huge cinnamon ferns (*Osmunda cinnamomea*). Energized by the archaeological discovery, faculty and staff at the middle school raised money for an eight-hundred-foot boardwalk through the wetland. Made of rot-resistant black locust, the path provides dry-footed access to an ecosystem that few people ever see. Signs along the boardwalk tell the story of the glacial lake and river, the native people, the thousands of years of peat buildup that converted the open marsh to a forested black spruce swamp, and the many cool plants and animals that make the wetland their home today. Middle school children now design and lead adventures in this "outdoor classroom."

Without a boardwalk, only a few hardy hunters and explorers make any direct use of swamps such as this one now. However, these ecosystems still provide huge benefits to modern-day humans, even though they may never set foot inside a black spruce bog or a red maple swamp. In addition to taking in carbon dioxide that might otherwise contribute to climate change, these forested wetlands help keep our streams flowing, thus maintaining aquatic life (Tiner 2003). The seepy swamps along small streams, high up in the watershed, are particularly important. These headwater streams and wetlands create an extensive network of water flows, covering a very large proportion of the watershed, allowing an enormous quantity of surface runoff to pass through them. Thunderous mountain storms and relentless spring snowmelts yield moisture to percolate through the dark swamp earth beneath the red maple, black gum, and yellow birch canopies. Weeks, months, or even years later, that water seeps out, coalescing into rivulets that finger their way downhill to the big rivers.

Water passing through these swamps gets subjected to the same filtering and chemical purification process described in chapter 2. Sediments are physically removed as the water slows down and is unable to carry the silt and fine sand it picked up from construction sites, forestry operations, road cuts, and driveways. Nitrogen, an important plant nutrient in soil, is also a potential pollutant in lakes and streams. Nitrogen is removed by chemical pathways that require some areas of organic soils with low oxygen and some with plenty of oxygen. Low spots will be full of water and will lack oxygen, and higher and drier spots will have plenty of oxygen; the combination converts harmful nitrogen in the water into pure nitrogen gas in the air (see chap. 3). The lumpy, bumpy forested wetlands such as red maple swamps create the perfect situation for this chemical process (Rheinhardt 2007). These same physical conditions can also create detoxifying processes for other chemicals. And because such a large quantity of flow passes through these many networks of headwater streams and wetlands, they play a more important role in water purification than the big mainstem rivers. Forested swamps can remove 80% or more of the nitrogen from farm fertilizers, stormwater runoff, and other sources (Tiner 2003). Clearly, the benefits of forested wetlands carry on from prehistoric times to today.

Riverside Swamps and Floodplain Forests

Author Catherine writes: The brown road sign depicts two stick figures paddling a canoe, marking an access point along the Connecticut River that is not meant for shiny extended-cab trucks towing bass boats on trailers. The narrow dirt road winds past long swathes of corn, already knee-high despite the cold, wet spring. The parking area sports a series of long mounds of silty, light-brown sediment curving around deep, trailer-size trenches, favoring four-wheel-drive trucks with canoes or kayaks strapped to the truck bed. For my low-riding, easily stuck economy car, finding a place to park here is an exercise in faith that the July heat has sucked enough water out of the ground to create a stable substrate. I seek to avoid the low, dark patches, which just last week contained puddles of water harboring gray tree frog tadpoles, and where now the fine muck threatens to mire my wheels. Finding a spot on the grassy edge, I exit on the side that doesn't put me into a patch of sting-ing nettles or poison ivy. Five steps out of the car and my feet are covered in a fine, flour-soft brownish-gray powder. The floodplain forest in summer.

Most people head for the river, seeking places to cast for largemouth bass, pike, or walleye, to launch a canoe, to picnic on the sandy shore. Arch-ing silver maples (*Acer saccharinum*), their skinny five-pointed leaves touch-ing the water, bear several rope swings, and the base of their trunks are lit-tered with the evidence of teenage partiers. Turning away from the river, I pass leaning cottonwoods (*Populus deltoides*), their bark deeply furrowed and bearing good-size holes, some of which contain old birds' nests. In late spring, cottonwood seeds float down through the air like a summer snow-storm, leaving drifts of fluff among the water-stained leaves on the ground. Side streams and backwater oxbows reflect the deep-green canopy of elms, willows, and maples in rusty-red and silty-brown water. The ground slopes down, slick greasy mud showing the hand-shaped prints of raccoons and the glossy shells of freshwater mussels. A mink runs along the far bank, hardly noticing my presence as it dashes under fallen logs and exposed tree roots. More distressed is a tiny black-and-orange warbler, an American redstart, bopping around the interior branches of a box elder, loudly expressing his displeasure over my trespass into his territory.

Floodplain forests throughout North America are rich from the soil below to the very tops of the trees, full of life. These forests are found along major river systems and their large tributaries — in the eastern and midwest-ern states of the United States, these big rivers include the Wabash, Illinois, Ohio, Susquehanna, Connecticut, Hudson, and, of course, the great Missis-sippi. Thousands of years of river flow has left dark, mineral-rich soils de-posited on top of the underlying alluvial layers of sand, silt, clay, and gravel.

March, April, and May bring high water levels, from spring rains and snow-melt. When large rivers flowed freely, without dams, levees, channels, or diversions, these spring flows inundated the lowlands around the river to a depth of five or even ten feet high. The muddy brown water moved slowly along the shallow slope of these big rivers, and fanned out broadly onto the floodplain, leaving behind loads of sediment, branches, leaves, and even whole trees. Testimony to the springtime floods shows up as waterlines on the trees, dark bark turning abruptly to light. But by summer, the waters have receded and the ground is bone dry.

All major rivers have this high water–low water cycle. In spring at Terre Haute, Indiana, for example, the Wabash River runs from only ten feet deep in a "normal" year, to as much as fourteen to sixteen feet in a wet year. At twelve feet, the river is full from bank to bank, and some low-lying fields will flood. At eighteen feet, farm fields and residences get inundated. At twenty-eight feet, a major catastrophic flood is in progress as water overruns highways as well as commercial and industrial properties, and residents begin to sandbag around their local school. In April 2013, the river crested at 27.4 feet, the highest level since 1958 (National Weather Service, n.d.). One hundred years earlier, the record flood was set at twenty-eight feet. Hundred-year flood events also took place in 1937, 1982, 1991, and 2005, certainly calling into question the concept of a "hundred-year flood" (based on statistics, such a flood is expected to take place only every one hundred years). The floodplain forests alongside these rivers absorb the high waters and allow the water to spread out, preventing flooding downstream and capturing pollutants. To understand how wetlands like this affect water levels and water chemistry, wetland scientists often install equipment such as monitoring wells to measure water levels and groundwater pressure, and to provide water samples.

Most wetlands have nice, soft organic soils or muck—relatively easy to dig through with a sharp shovel or spiral auger. Not so on the river edges near riparian forests, as Andy Cole of Pennsylvania State University found while attempting to install some shallow wells alongside the Little Juniata River in Tipton, Pennsylvania. Here, just below the initial layers of fine silt and mud, Andy encountered large cobbles, sand, and gravel. No shovel would slice through these rocks, so rather than painstakingly digging each one out to create the five-foot-deep hole needed for the monitoring well, Andy decided to bring in the heavy artillery. He headed to the local rental place to borrow a jackhammer—along with the huge gas-powered generator needed to power it. "Amazingly, they didn't ask me if I knew how to use the equipment, and I didn't, really, but I figured it out. What I didn't realize was that the jackhammer wouldn't split the cobbles—they just slid away from it. So we got nowhere." Meanwhile, the huge generator had sunk axel-deep in the mud and required a local tow-truck operator to winch it out. "I was terrified to bring it back to the rental place—it was dented and covered in mud—but they didn't bat an eye," he remembers. "The whole thing was a disaster. My crew was looking on, thinking, 'We *thought* he knew what he was doing!' After all that, I found that it was easier to dig the hole the old-fashioned way—with a shovel and my bare hands."

Fortunately, Andy and his coworkers were able to get the information they sought. His studies, along with others, have shown that wetlands along the mainstem of major rivers are among the driest of the various wetland types, with water in the plant root zone less than half the year (Cole and Brooks 2000). These dry periods provide ample space to absorb the floods coming in from snowmelt and spring rains in the early part of the growing season.

Sanctuary, II: A Bouquet of Birds

Paddling his one-person canoe in the backwater sloughs of the Mississippi River in spring, Minnesota birdwatcher and writer Richie Swanson knows he needs to avoid the current of the main channel to stay out of trouble. "The water is maintained [by the locks and dams downstream] in the backwater forests at about five feet deep. But way over in western Minnesota they will get some nine inch rainstorm, and ten days later that water will make its way down the Minnesota River and into the Mississippi, and the water will rise six or eight feet. Spring to spring, you never know what the water level will be—it could be twelve inches or it could be fifteen feet," he explains.

Historically, much of the floodplain forest along the Mississippi would have dried out regularly by late August or September; but since the 1930s, when locks and dams were installed at regular intervals along the river, that doesn't happen. Now it's wet almost all the time, so the floodplain trees can't reproduce—the seedlings drown in the deeper waters. Silver maples are the only tree that can survive the new hydrologic regime. Invasive reed canary grass moves in quickly to any sunlit opening, shading out any tree seedlings that had the good luck and gumption to sprout (see chap. 2 for more about reed canary grass). What used to be floodplain forest is now a grassy open marsh in many areas.

Richie spends a lot of his time exploring Aghaming Park, a beautiful piece of the floodplain forest along the Mississippi River, owned by the City of Monona, Minnesota (but located across the river in Wisconsin). After reading *Where Have All the Birds Gone?*, by John Terborgh, he decided to do something to help bird conservation by conducting a breeding-bird census each year in the sloughs. "There's a lot of life in the floodplain forest," Richie laughs. "Most people's first impression of the floodplain forest is poison ivy and nettles and mosquitoes. They think of it as hostile grounds, but it is full of life!" His devotion to this place, and these birds, is profoundly moving.

In a breeding-bird census, trained volunteers go out to specific areas and identify all the birds they see or hear in a set amount of time. "Early on a June morning, there would be songs everywhere—more than a guy could write down in a minute. I mapped the area into a grid of fifty-meter squares, and as I crept through each block, a catbird would be bursting out of a bush on one side of me, or a robin would be dive-bombing my head, maybe a great crested flycatcher would be calling from a treetop. In spring, the buds are just coming out, the tops of the trees are brushy with the wine color of maple buds, and the yellow of the cottonwood buds. You see mink, and opossum, and beaver and otter. Here on the Upper Mississippi there are these limestone and sandstone bluffs, like palisades, bordering both the Minnesota and Wisconsin sides. It is a river system squeezed between highways and railroads, so it is hardly pristine. But you can easily get to a place where it's hard to see a house. If you walk a half hour into the woods, you've left everybody behind."

Big rivers like the Mississippi have been dammed, dug, diverted, and channelized, and the wetlands alongside them have been drained, farmed, filled, and degraded. The areas that remain, like Aghaming Park and the nearby Upper Mississippi River National Wildlife Refuge, become all the more critical for wildlife. Richie describes one of his favorite species, the prothonotary warbler (*Protonotaria citrea*) a tiny buttercup-yellow bird with a blue-gray back and wings. "The prothonotary warbler provides the comic relief for all the bad things that are happening. You are in this swamp that's full of mud; it is always dark because it is in shadow. There's this wet, dark stump that looks rotten and worthless, and all of a sudden out of a hole comes this bird that is just impossibly bright. It sings a very pronounced, ostentatious song. The male displays at the hole and flashes its white tail feathers five or six times in a couple minutes; he will pop into the hole to lure the female inside. Sometimes he will have a piece of moss in his bill—he claims the hole and the center of his territory with that piece of moss." After the male starts building the nest, his mate will finish up the construction, clacking her bill at intruders while she works. Soon she will settle in and lay four to five brown-splotched white eggs.

Birds that build their nests in holes are called cavity nesters, and they depend on the presence of decaying trees to provide the crevices needed for nests. The flashy-yellow prothonotary stages territorial battles for prime space with house wrens and flying squirrels. With the decline in floodplain forests, prothonotary warblers have been found to explore other nesting options—including nest boxes, cement blocks, old hornet nests, a mailbox, a glass jar, and a teakettle.

Warblers such as the prothonotary are small, insect-eating birds, often so colorful that a group of them has been called a bouquet. All of them migrate to tropical areas of Central and South America in cold weather, and thus are vulnerable to habitat destruction in both their winter and summer homes. Only 10% of the prothonotary warbler's breeding habitat in the bottom-land swamps remains, while as much as 70% of the mangrove swamps that it chooses for wintering in Colombia and Ecuador have been destroyed. Its population has declined by 42% since 1966 (Petit 1999), and several scientific groups have documented its continued decline (Cornell University Laboratory of Ornithology 2016; Sauer et al. 2017). "The Upper Mississippi River area sustains about 20% of the global population, so it is really important," Richie notes.

Richie is also enamored by another small bird: the handsome blue-and-black cerulean warbler (*Dendroica cerulea*), which hides itself away from easy viewing high up in the top of the cottonwoods. "It's this tiny little bird among the green glittering leaves. They have become really rare. At Aghaming, when I did the point counts, when I stepped out of my Toyota I often heard an instant, rhythmic song from the cerulean, like it was saying 'Save me, save me, here I am.'

"There are so many species of interest here [in the Upper Mississippi]. There's the rusty blackbirds with their clamorous noise and their dramatic swarms of hundreds of birds. And red-shouldered hawks — I spend a lot time trying to find their nests to make sure they don't get disturbed. Whenever I find a nest, I think, 'Okay, I'm in a sacred place.' I do objective science for the bird counts but at the same time I love them all as I watch them, they are so full of energy and life. They all have personalities. All that effort each bird puts [into] migration and nesting — and if just one bird survives and returns, it's a success. Long ago I learned to sit quietly and watch what goes on."

Bats in the Farm Belt

To see some of the other interesting inhabitants of the floodplain forest, you also need to sit just quietly and watch. Try watching at night — for bats. We don't generally think about bats when we think about wetlands, but they are there, feeding on insects at night and roosting (hiding and sleeping) in hollow trees and under scaly tree bark during the day. In fact, sometimes one tree will host hundreds of female bats and their babies every night, in what is called a maternal roost. Wildlife biologist Jacques Veilleux of Franklin Pierce University tells how he finds these roosts. "At night, we string out large nets — nine meters wide by about nine meters high. The bats fly into the nets, and we catch and measure them, put a radio tracker on them, and follow them to a tree. Then we go back to the tree at dusk and do what is called an emergence count — just watch them come out and count them. A lot of trees, especially silver maples, have huge hollow cavities, sometimes from bottom to top. This creates roosting habitat, particularly for the evening bat [*Nycticeius humeralis*] in bottomland hardwoods in Indiana. We'd find upwards of three hundred–plus females in one maternity colony." These large, dead trees, with rotting cores or exfoliating bark, are important for many species of bats, including the endangered Indiana bat (*Myotis sodalis*) (Carter 2006).

Jacques explains why there are so many bats in these forested wetlands: "Bats like wetlands because there are a lot of insects, and a lot of water. These maternal roosts are really hot, baking in the sun all day; the first thing the bats do each evening is go and get a drink. Wetlands are always massive foraging habitat for bats, even if there are no big hollow trees."

Most bat species need wetlands, but bats cannot live on wetlands alone. Many of them also need upland forests to find specific prey items. Or, some species do the reverse: they bed down in large old trees in uplands, but dusk finds them heading to open water or wetland habitat for dinner and drinks. This includes the Indiana bat, which is particularly dependent on the bottomland hardwood forests associated with large river systems in the Midwest (Carter 2006). Wildlife in all wooded wetlands seem to share interdependence with the adjacent uplands (see chap. 6). Many of the bird species that live in the wet forest or swamp also need the upland forest nearby for some part of their life cycle (Riffell, Burton, and Murphy 2006).

Along the tributaries of the Wabash River in Indiana, where Jacques conducted his thesis research, fields of corn and soybeans have been planted where the floodplain forests and bottomland hardwoods used to flourish. In many areas, a narrow ribbon of this forest remains along the stream corridors, serving as important buffers for floods, filters for fertilizer-laden runoff, and excellent habitat for wildlife. The farmers who own these lowland swamps along the river may or may not appreciate the bats, which eat a lot of the crop-destroying insects; but they sure don't like flooded fields. "Every year the Wabash would rise, fill up the creek—there would be no more creek. It would fill up the woods and literally we would be hiking in the woods in waist-deep water," Jacques says. "Huge carp would swim right into the forest, the water was so deep; you'd see the weeds shifting, then whack—they'd hit your boots. By mid-May or early June, the waters would recede to where it was back down to a reasonable level, to the top of your rubber boots or lower."

Decades earlier, the local farmers had attempted to drain the wetlands by digging a ditch to get the water to flow out of the forest and adjacent fields faster in the spring. "One of our highly productive bat-capture sites was along one of these straight ditches. On either side of the water, there was a beautiful, thick-forested canopy of silver maple, green ash, American elm [*Ulmus americana*], red elm [also called slippery elm, *Ulmus rubra*], and sycamore [*Platanus occidentalis*]. In the third year of my study, we showed up at this site and saw that the farmers had come in with equipment, and they just dredged it out—no permit or anything. They wanted to get the water out, to let their fields dry faster. My advisor went down there, and he was blown away by what he saw. He reported it to the EPA. These farmers, that I still had to work with, were fined over $100,000. For the next two years, I had to go down there; they knew it was me that discovered what they did. So one day, I am by myself, in a university minivan, parked in the middle of nowhere in the bottomlands. The van is all dusty. I am out there in the middle of the night, netting until two in the morning, and when I come back to the van, I see, written in the dust on the back window, two big eyes and the words 'The Klan is watching YOU'! I was terrified, I thought for sure there were dudes in back of a tree waiting to do me in. So I opened the back hatch of the van, dove in, and just drove away. Left all the equipment behind, figured I could come back in the daylight to pick it up."

"Some of the farmers were really nice—they still let me go out on the land. Large tracts of land were owned by several families—they'd just let us go do whatever we wanted. They didn't care, although they started to care after this event!" Although Jacques finished his work safely, the incident stays with him after more than twenty years.

Siren Song of the Bottomlands

A few states away from Jacques's bat study in Indiana, wetland ecologist Frank Nelson spends his days piecing together the ecological puzzles of the Mingo River basin in southeastern Missouri. The braided channel of the Mississippi River historically ran through this area, before changing pathways and leaving behind thousands of acres of wet bottomlands, which were kept watered by flow from the Castor River. In 1918, a diversion channel was built to shunt the water off these bottomlands thirty miles west, to the Mississippi River. Eventually, over six thousand miles of drainage ditches were built, drying out the lowlands for farming.

In the 1940s and '50s, some of the wettest areas became part of the Mingo National Wildlife Refuge and the Duck Creek Conservation Area. "Much of the culture around here is all about hunting," Frank explains. "So the managers emphasized habitat for waterfowl." They created wet basins containing open-water marsh, lakes, and forested swampland known as "green tree reservoirs" in the Duck Creek Conservation Area, all of which are important habitat for pintails, teal, mallards, shovelers, and other ducks and geese. "Back fifty or sixty years ago, when these areas were developed, the engineers didn't think about how water flowed across the landscape; they just put some levees down, made it a rectangle, and they put the water-control structure in the lowest place. But the water just sits at the lowest spot and doesn't get the water off of the trees." So the trees drown, and the creatures that depend on the forested wetlands lose their habitat. Part of Frank's job has been to improve the water management so that a greater diversity of natural communities can survive. Using handheld GPS units and lots of swamp walking, Frank and his colleagues mapped out the detailed contours and curvatures of the land, trying to figure out where the water used to flow and how changes in water levels would affect the plants and animals.

Very small differences in elevation—less than a foot—make a big difference in the type of forest that grows in these lowlands. "Because the rivers have shifted and moved over time, you have these small landforms, these little ridges; they may only be a foot or two, but the forest community has distinct changes along these ridges. There may be upland species like cherrybark oak [*Quercus pagoda*] and willow oaks [*Quercus phellos*], water hickories [*Carya aquatica*], even pawpaw [*Asimina triloba*] and hawthorn [*Crataegus* spp.]."

Walk a few inches downslope here (as well as in other southern sections of the midwestern and mid-Atlantic states) and the true wetland trees take over, such as the bald cypress (*Taxodium distichum*), a coniferous tree with rows of short needles and incredibly rot-resistant wood. Like the tamarack, this conifer sheds its needles every year, making it a deciduous conifer, which sounds a lot like an oxymoron but is not. Unlike pine and spruce, some cone-bearing trees drop their needles all at once. Thus the bald cypress is depilated each fall. Alongside the cypress, tupelo trees are a common companion. Frank describes what it is like to visit these cypress-tupelo swamps in Missouri, at the northernmost edge of their range: "The tupelo turn yellow in the fall, and if you can catch it just right, those leaves have started to fall and scatter across the water. The cypress needles provide some cinnamon-brown contrast, and the light just filtering through the canopy reflects off that golden-brown surface—it's just spectacular."

In response to the constant high water, the bottom part of the cypress trunk flares out to as much as three times the width of the upper trunk. The cypress also produces "knees," which are woody-gray, vertical protrusions of the tree's roots poking up out of the water. Early suspicions that these knees transport oxygen to the waterlogged roots were rejected due to lack of air-transport structures such as lenticels or aerenchyma. Recently, however, a team of researchers measured oxygen pressure in a root when its knees were submerged and when they were in open air, and found that these strange structures do in fact bring oxygen into the tree's roots, thus helping the bald cypress survive extended flooding (C. Martin and Francke 2015).

Probing the dark waters of the cypress-tupelo swamps, especially at night, one might come across a long, snake-like beast with frilly gills around its head—a miniature Loch Ness monster. No need to worry about being bitten; it is a toothless salamander called the lesser siren (*Siren intermedia*). Despite its name, it has no beckoning siren song, just a yelp or a click when captured. However, since no other salamanders make any noise at all, these calls alone may justify its alluring title.

As part of his responsibilities, Frank Nelson and his team conduct regular sampling of the fish and amphibians using mini fyke nets, which are made of circular or rectangular metal rings, arranged in a series of descending sizes, all connected with nylon webbing. "In 2011, which was an amazing flood out here in the Midwest—over twenty inches of rain in ten days—the Mingo basin was flooded for over a month, so we went out and sampled; we wanted to find out who was using this habitat. We set the nets out in tandem, always leaving part of the net out of the water for the turtles and others who need to come up for air," Frank explains. "We would take a boat out to check the nets, it was kind of like Christmas—you never knew what you were going find! In a month we caught over forty-six species of fish. One time, in one net, it was so full we had probably over two thousand tadpoles, hundreds of crayfish, sunfish of a variety of sizes, pickerel, a bowfin about twenty inches. That bowfin was a lesson in gluttony—he had five different fish in his mouth, and one of those fish had a fish in its mouth. And we also caught several big sirens, some over one and a half feet long. At first, you'd see this long, linear body, and you weren't sure if it was a snake or a siren. The sirens were really hard to handle—very slimy and lots of muscle, always trying to wriggle out of your hands. One time I saw a great blue heron messing around with a siren in its mouth. I thought it was a snake until I saw the arms. The heron was struggling with it as much as we do!" Catching a siren in the middle of the day was a real feat for the heron, because sirens are active mostly at night in order to avoid that exact fate. "The next year was a drought, and [to survive] the sirens will secrete this slimy mucus to make a parchment-like cocoon in the mud." They then sleep in a low-metabolic state called aestivation. "They are in different locations from one year to the next, the bottomland swamps as well as in the basins and reservoirs."

Entering a forest on the drier end of the forested wetland spectrum, explorers are unlikely to run across sirens, but they may encounter (depending on the location) an overstory dominated by white oak (*Quercus alba*), American beech, (*Fagus grandifolia*), Carolina ash (*Fraxinus caroliniana*), overcup oak (*Quercus lyrata*) and sweet gum (*Liquidambar styraciflua*). Sweet gum, with its lovely star-shaped leaves and its painfully spiky seed pods, was recently discovered to be an important source of shikimic acid. This molecule is the critical ingredient for Tamiflu, a drug that fights influenza by preventing the flu virus from reproducing (E. Martin et al. 2010). Historically, sweet gum sap was long recognized as an important medical gum, used by Native Americans and early settlers for the treatment of wounds and diarrhea and as a sedative (Moerman 1998). "Overcup oak are really neat too," Frank says. "There's a lichen community on their bark that makes them almost teal, sometimes white. The corky caps on the overcup acorns are basically like little life vests—you will see rafts of them floating to the edge, where they establish along the highest waterline. And the overcup acorns are so huge, a squirrel could feed on one for days."

Ironically, many of the efforts to drain wetlands to create drier conditions in an average year will increase flooding during a wet year. Because wetlands store floodwater, people have created channels to drain the water out, shunting stormwater right through the floodplain into the river, hastening the arrival of high water downstream. In addition, wide floodplains provide a lot of what is called bank storage—not money stored in a financial institution, but a commodity far more vital: water. Groundwater runs high in spring, moving slowly from the wet areas uphill into the valleys below. Wide floodplains—especially those with sandy soils and adjacent to steep slopes—will store groundwater, slowly releasing it over weeks or even years (Tiner 2003). When this groundwater flow is linked to a healthy network of streamside wetlands all the way to the top of the watershed, the river is buffered against droughts as well as floods.

"These lowland wetlands are one of the habitats that society has given a bad name," Frank laments. "We've lost so much out here in southeast Missouri. We are just trying to restore things to provide the same functions that they may have provided historically. We are looking for ways to make connections among the pieces we have left. We derive many of the values and knowledge from the world we experience. I remember one spring, during a family outing to a restored wetland area. My kids were doing what kids do best, running around, not really aware of their surroundings, but taking it in all the same. I looked at the redesigned marsh area, and I saw water flowing out of the banks of the meandering channel. I saw abundant life utilizing the habitat. And I saw another generation experiencing nature, making memories, and formulating values. In moments like these, I am reminded why conservation is important."

CHAPTER 6

Vernal Pools: Believing in Wetlands That Aren't Always There

> The woodland depression that cradles the vernal pool seems so
> abandoned without its water. In its present empty state, I cannot
> bring into clear focus a vision of what the pool was like little more
> than sixty days ago. At the same time, there is an air of patient
> waiting in this hollow, an evocative sense of tempered expectation,
> that I find nowhere else in the surrounding forest.
> —DAVID M. CARROLL, *Swampwalker's Journal: A Wetlands Year*

Scott Jackson had been hauling critters out of pit traps for several days during a rainy spell in Granville, Massachusetts, when he found an eastern spadefoot toad (*Scaphiopus holbrookii*), a threatened species in the state. One spadefoot toad is pretty exciting, given that at the time there were no known populations west of the Connecticut River in Massachusetts; but each day brought another toad . . . and then a whopping three toads in one go. Well, this really was something. So Scott did what any curious herpetologist would do: he went looking for spadefoot toad central—their core habitat. Scott's search for the spadefoot's home ground led him to an unexpected wetland treasure, a surprising discovery, and a lifelong memory.

The vernal pool in Granville is a well-known amphibian hotspot, a spectacular breeding site of six and a half acres boasting thousands of salamanders and countless peepers, wood frogs, and tree frogs. The site is also what you might expect of a classic vernal pool—a low-lying spot in the woods that fills with water in the spring and dries out as summer progresses. Scott, an associate extension professor at the Department of Environmental Conservation, University of Massachusetts, was surveying the amphibious bounty of the Granville vernal pool when he came across the spadefoot toad in one of the pit traps (*pit trap* is the technical term for a hand-dug hole in the ground into which amphibians and reptiles fall and wait to be rescued and counted by a herpetologist). Only two to three inches long, the spadefoot is unique among toads, sporting sharp-edged, sickle-shaped spades on its hind feet and golden, cat-like eyes with vertically slit pupils (Massachusetts Division of Fisheries and Wildlife 2015).

The spadefoot toad is a rare find in the Northeast and in parts of its range that extend west because it absolutely depends on vernal pool habitat for breeding. It spends a great deal of time buried underground and comes out only at night, which also thwarts detection. The toad uses those spades on its hind feet to dig a hibernation burrow, excavating backward, up to eight feet down. During dry spells, the toad will remain in its burrow, having secreted a fluid that hardens the earth walls of its den, capturing enough moisture to survive (McCormack, n.d.). If you happen to be lucky enough to catch one, beware: those secretions may elicit an allergic reaction much like an allergic reaction to a cat (remember its eyes)—violent sneezing and watery eyes (Connecticut Department of Environmental Protection, n.d.).

Returning to the forest and pool at night to listen and look for a breeding population of the toads, Scott stepped out of his truck to the sound of a flock of bleating sheep: spadefoot males, calling for mates. Just what he had hoped to find. But the cacophony was not coming from the interior of the wood, where the vernal pool lay. Following the noise led Scott *away* from the vernal pool down the road to a flooded, uncut field of grass and clover. It was a warm summer evening, the sky full of stars and the flooded field running over with the deafening, desperate riot of calling amphibians. As he waded barefoot and waist-deep into the perfectly clear water, Scott could see with the aid of his headlamp hundreds of spadefoot toads and gray tree frogs nestled in amid the vegetation. This qualifies as one of those moments when the learned biologist succumbs to muttering, repeatedly, "This is so cool."

So cool and so beautiful, he brought his wife out the next night to share the magic, wading barefoot among the clover and toads.

A drop in barometric pressure triggers spadefoot breeding on warm, rainy nights, and the party was still in full swing while the field remained flooded. As Scott watched, the male toads inflated like tiny balloons, their diminutive limbs sticking out from a central bubble; and when they called, they tipped up and down, flashing their white bellies. If startled by Scott's movements, they made a dive for safety, but being so buoyant they needed to grasp on to a plant to hold them under the water. When Scott was still and the scare over, they simply let go and bobbed back to the surface. Left undisturbed, the male eventually fertilizes the eggs as the female lays them in tenuous strings, draping them over the vegetation. A female spadefoot will lay up to 2,500 eggs and then exit the party, leaving the male to try his luck with another in the crowd. The temporary nature of the pool demands a quick life cycle, so the eggs will typically hatch in two to four days and the tadpole to toad transformation can take place as quickly as sixteen days if close to the end of the breeding season, or a more leisurely forty-eight days if they get started earlier in the spring (McCormack, n.d.).

The field-turned-toad-bacchanal was a serendipitous find, an unknown and unusual amphibian hotspot likely to be temporary should it be planted to a different crop or paved for new housing. The landowner needed tracking down, so Scott grabbed a few toads for props and went knocking on neighborhood doors. Imagine carrying around your preferred specimen and introducing it to a variety of strangers on their doorsteps. You're likely to receive a better welcome than the politicians and issue canvassers, and toads are much more interesting than petitions. Having spent years studying salamanders at the vernal pool breeding site, Scott was a familiar sight in the area and found people interested and amused as he made the rounds with his captured amphibian curiosity. After about four houses, Scott found the farmer who owned the hayfield but who had no knowledge of its sporadic residents. Wary at first, the farmer warmed after seeing the toads in the bucket, and surprisingly, he was quite pleased to know that they bred in his field. Intrigued, the farmer ended up taking Scott on a tour of his farm, pointing out all the vernal pools that were previously unmapped.

A new spadefoot spot is exciting enough to share with colleagues, so Scott brought the news to his coworkers at UMass Amherst. Like most university research spaces, science departments are a warren of labs, cubicles, and offices, but natural history spaces are often full of specimens or parts of specimens: snake skins draped over lamps, shelves of jarred salamanders and frogs, mammal skeletons serving as decorative sculptures. Behind the lab benches and desks are people who study all manner of creatures and plants, and they are bound to share in your enthusiasm. On this particular day Scott found Doug Smith, whose passion is a little-known creature called a clam shrimp. Clam shrimp really do look like tiny shrimp trapped in clamshells, their minute legs sticking out, but they are not mollusks. In this case, the organism in question is a tiny crustacean—a relative of crab and shrimp—enclosed in translucent, shell-like valves just a quarter inch or so in length. They live exclusively in vernal pools, and there are several different species. Doug was keen to find one particular species. So, while duly enthusiastic about Scott's discovery of a spadefoot toad breeding spot, Doug did make a plug for clam shrimp and asked Scott to keep an eye out for the aquatic creatures — "the big ones, not the tiny ones you see everywhere."

Thus the next time Scott visited the hayfield, he filled a bucket with "big" clam shrimp, which clouded the water with their numbers. Chore accomplished, he took the bucket to Doug, who picked his jaw up off the floor after looking into the bucket. "But . . . but there are only three occurrences of that clam shrimp in North America—where did you get those?" "Well," says Scott, "you told me to get clam shrimp from the spadefoot pond, so I got the clam shrimp." Turns out, those clam shrimp were American clam shrimp (*Limnadia lenticularis*) and known to exist in just three counties in Massachusetts and three in South Carolina, plus one population in Georgia. Doug had been looking in vain for these clam shrimp, and Scott had casually turned up with three or four hundred in a bucket. It also turns out, upon closer inspection, there was yet another rare clam shrimp in the flooded field, one that had never been documented in North America: Agassiz's clam shrimp (*Eulimnadia agassizii*). Three rare species in one humble field of clover and grass.

While spadefoots like to breed in open fields such as the one Scott Jackson describes, it is not common to find such undrained agricultural areas. The Massachusetts Division of Fisheries and Wildlife now owns the land and, on Scott's advice, that same farmer still manages the land as he always has.

The vernal pools in which spadefoots and other amphibians breed are considered important by many natural resource conservation agencies, and some are even given protected status. But vernal pools are temporary wetlands, appearing during much of the year as simply an empty space among the maple, alder, and buttonbush. Few people realize that, come spring, these empty spaces will fill up with water and burst forth with salamanders, newts, frogs, and fairy shrimp.

The Race against Time: A Year in the Life of a Woodland Vernal Pool

Ephemeral wetlands are just that — ephemeral. People who walk about in the late-summer woods may not detect what Paul Zedler calls the essential "poolness" of a particular depression in the landscape (2003). But it is the water's disappearing act that permits a set of unique species to exist. The seasonal woodland pools that we call vernal pools are mostly fishless — the seasonality creating a fishless environment that in turn establishes suitable habitat for the eggs and larvae of amphibians and reptiles. Though lacking any piscine predators, the temporary nature of the pool brings its own challenges.

Come winter, the pool is just a space among the trees, a little lower than the surrounding land surface and likely filled with leaf detritus from the fall. Under all that litter is incredible dormant diversity: diapausing mollusks, quiescent worms, larval salamanders, and a freeze-dried egg bank of crustaceans. Surrounding the depression, underfoot, are frozen frogs and hibernating salamanders. While the toads and salamanders will hibernate deep underground to avoid freezing, four frogs common to the Northeast and the upper Midwest actually freeze just under the litter until spring thaw. Spring peepers (*Pseudacris crucifer*), gray tree frogs (*Hyla versicolor*), chorus frogs (*Pseudacris triseriata*), and wood frogs (*Rana sylvatica*) all tolerate ice in their extracellular spaces; 35–65% of the water in their little bodies freezes, their breathing, blood flow, and heartbeat coming slowly to a stop (Storey 1990).

> Frogs of various colours are numerous in those parts as far North as the latitude 61°. They always frequent the margins of lakes, ponds, rivers, and swamps: and as the Winter approaches, they burrow under the moss, at a

considerable distance from the water, where they remain in a frozen state till the Spring. I have frequently seen them dug up with the moss, (when pitching tents in Winter,) frozen as hard as ice; in which state the legs are as easily broken off as a pipe-stem, without giving the least sensation to the animal; but by wrapping them up in warm skins, and exposing them to a slow fire, they soon recover life, and the mutilated animal gains its usual activity; but if they are permitted to freeze again, they are past all recovery, and are never more known to come to life. (Hearne and Tyrrell 1911)

Late winter into early spring brings snowmelt and rain to depressions in the forested landscape. As the temperature warms, the slimy and the slow thaw out from winter dormancy and begin their overland trek to the pooled water in which they were born. Waiting for them are hatching, pink-orange fairy shrimp (*Eubranchipus* spp.) who spent the winter as eggs, freeze-dried in the bed of the dry basin. Fingernail clams (*Sphaerium occidentale*) that burrowed down into the mud the previous summer now break dormancy to once again filter the detritus-laden water. Walking in the springtime woods, you will likely hear the vernal pool before seeing it. The wood frogs are the first frogs out of hibernation. Having spent the winter frozen beneath the leaf litter of the upland forest surrounding the pool basin, they head for the water during the first warm spring rains. This small brown frog with the bandit mask (hence the nickname "robber frog") makes a calm quacking noise when it calls. Consequently, a pool full of wood frogs can sound somewhat like a contentedly crowded chicken coop or mallard duck pond. During breeding, the female wood frog will deposit up to a thousand eggs in one spherical mass attached to some submerged woody debris, shrub, or remnant of last year's perennial herbaceous stem. The male, clasped to her back, fertilizes the eggs as they are laid. In some vernal pools, the wood frogs lay eggs together, forming a large communal raft, making the surface of the water the texture of jelly.

Also rustling out from under the leaf litter at spring are mole salamanders (family Ambystomatidae) who have spent the winter in burrows repurposed from small mammal lodgings. Rather than blend into the forest floor, the spotted salamander (*Ambystoma maculatum*) accessorizes its blueblack skin with large bright yellow or orange spots that advertise noxious excretions to potential predators. More subtly accessorized is the blue-spotted salamander (*Ambystoma laterale*), which sports a tasteful spray of smaller blue dots along its dark body. The Massachusetts Division of Fisheries and Wildlife (2016) fact sheet describes the salamander's color pattern as reminiscent of antique blue-enamel dishware. This is not an intimidating description and probably not enough warning, given that the blue-spotted salamander will lift its tail when threatened and then squirt a milky toxin into its attacker's mouth. Not sufficiently protected by toxins, both salamander species travel only during the rains of warm spring evenings—the journey made under cover of darkness to skirt abundant daytime predators, in the rain to keep their skin moist, and when temperatures are warmer to avoid freezing midstride.

Salamanders' travels take them as far as a mile, sometimes over patches of snow, most likely to the same pool in which they hatched. The males gather in large clusters, called salamander congresses, once they reach the water. They writhe around in a jumble of tails and limbs and then deposit small, white packets of sperm, called spermatophores, on leaves and other debris in the pool. A male blue-spotted salamander will attempt to lure a female to his spermatophore with rubbing, nudges, and an embrace, which sounds a whole lot more romantic than the scientific term for the "embrace": amplexus. If the female follows, she will pick up his spermatophore with her cloaca (which is her outlet for the urinary, intestinal, and genital tracts). It's basically the same routine for the spotted salamander (the one with the yellow spots) except, eschewing the embrace, the male courts females by swimming upward from the bottom of pool while weaving his head left and right—a dance move sure to bring the ladies to his side.

Each spotted salamander female will lay one to three egg masses, each containing fifteen to one hundred eggs. Individual eggs are surrounded by a milky-white gelatin, and the entire mass is surrounded by clear jelly—the whole thing looking very much like a scoop of tapioca attached to a submerged stick. Again, the blue-spotted salamander egg mass is more understated, with just one or a few eggs per mass protected by a clear, runny jelly, and often deposited on leaf litter in addition to submerged woody debris. Blue-spotted salamander eggs, as far as we know, also lack a particularly spectacular skin accessory found in the more common spotted salamander—algae. Scientists speculate that the algae, which can be seen surrounding the embryo in its egg case, are actually incorporated into the developing embryo's cells. The algae may have even originated in the mother, who passes the algae to the eggs as they are laid. As you might guess, the conjecture is that the algae feed off the waste produced by the salamander's skin cells (nitrogen and phosphorous) and in turn produce oxygen, increasing the embryo's chance of developing successfully (Kerney 2011). Bright yellow polka dots and algae — now, that's high fashion!

Meanwhile, while all the amphibian breeding and egg laying is in full swing; the reptiles are advancing on the vernal pools, as are the predacious beetles and dragonflies.

Spring turns to summer and the pool warms and shrinks. Great herds of wood frog tadpoles fill the pools of the early to midsummer woodlands. At one particular pool in a southern New Hampshire hemlock-and-red-maple forest, the water is still cold but perfectly clear; every detail of the white pine needles and deciduous leaves littering the bottom is visible. The tadpoles appear black above the rusty colored detritus and scatter upon approach, piloting their fat little bodies with their sleek but temporary tails. The faster the pool dries, the faster these tadpoles will metamorphose into juvenile frogs. The salamanders have not yet hatched, so the frog tadpoles feast on the algae coating the egg masses of their fellow pool inhabitants, occasionally ingesting the defenseless embryos as well (Petranka, Rushlow, and Hopey 1998) — in addition to smaller bullfrog tadpoles, slugs, beetles, and snails (Hunter, Albright, and Arbuckle 1992). In turn, wood frog tadpoles are prey to diving beetles and adult salamanders. Caddis fly adults emerge, and a whole host of larval and adult aquatic beetles (Order: Coleoptera) and bugs (Order: Hemiptera), and dragonfly nymphs (Order: Odonata) are on the prowl for other insects, tadpoles, and one another.

By the time the feather-gilled salamander larvae finally hatch, the pool is a living, writhing buffet for tenants and visitors. The snapping, Blanding's, and spotted turtles, the water snakes and the water birds, the mink, shrews, moles, and raccoons come to feast on all manner of larval and adult pond dwellers (Hunter, Albright, and Arbuckle 1992; Paton 2005). In the northeastern United States, three of fifteen species of snakes use seasonal pools for foraging or basking, and six of twelve species of turtles depend on these pockets of water for at least one stage of their lives (Paton 2005). The turtles are here not only for food but also to mate, although they will lay their eggs at some distance from the wetland.

Insects make up the greater portion of total diversity and biomass in the vernal pool, but the one insect so many people equate with the now warm standing water is the mosquito. At least thirty species of mosquito are identified with vernal pools (Colburn 2004), but the mosquitoes we encounter when we visit a vernal pool represent, incredibly, only 1% of the potential population of mosquitoes maturing out of that pool. The remaining larvae constitute a major food supply: the nonvegetative base of the food web, supporting beetles, caddisflies, bugs, nymphs, newts, frogs, salamander larvae — just about anything that makes a living in the vernal pool. Consequently, mosquitoes really prefer a pool without all the tadpoles and salamander larvae, depositing fewer eggs in pools with greater numbers of these insect predators (Rubbo et al. 2011).

If you can stand still long enough while the mosquitoes swarm, take a close look at the moss-covered twigs in the pool. You just might see something altogether unexpected and odd: tentacles. Generally speaking, moss does not have tentacles, but moss animals do. Bryozoans, commonly called moss animals, are colonial creatures that bring to mind marine corals extending their tentacles to filter feed. Most moss animals are marine, but one class, the Phylactolaemata, is exclusively freshwater and can resemble gelatinous globes, fuzzy caterpillars, or moss (D. Smith 1992; Burne 2013). Next time, that fuzzy green stick just might be worth a closer look.

Time's Up—Everyone Out of the Pool!

At some point during the summer, the salamander larvae lose their gills and the tadpole replaces a tail with legs. They then leave their natal home to make their way in the surrounding forest. Wood frogs leave the pool by the hundreds. With every step you take around the pool, dozens of wood frog metamorphs (tiny frogs) leap to safety. Safety is relative, of course, as the exodus from pool to woods is a convenient food delivery system for upland creatures like snakes, raccoons, skunks, foxes, and birds (Hunter, Albright, and Arbuckle 1992). Many vernal pool denizens abandon their shrinking habitat for deeper pools or head for the more permanent waters of a swamp or shallow marsh. Any amphibian not making the critical transition from aquatic to terrestrial dwelling creature is out of luck by the end of the summer as the pool dries up. The all-you-can-eat buffet for larger and more mobile creatures is closed and the reptiles move on to spend the rest of the summer in the uplands or find deeper waters in which to continue feasting.

Not all is lost when the water is gone. The invertebrates—the mollusks, insects, crustaceans, and worms—are preparing for the dry time. Oligochaetes (aquatic worms) are most abundant in the wet soil at the end of the season. They, like the fairy shrimp and clams, will go dormant when the pool is dry and overwinter in the sediment. There are quite a variety of survival methods employed by aquatic creatures that remain loyal to the once wet location on the forest floor. Oligochaetes can cover themselves in protective mucus, fragment as cysts, or leave their eggs to carry on in desiccation-proof cocoons (Colburn, Weeks, and Reed 2007). Crustaceans—the group that includes fairy and clam shrimp and water fleas (Order: Anomola)—also contribute desiccation-resistant eggs to the "egg bank" of the pool (Colburn, Weeks, and Reed 2007). Fingernail clams will stop growing, burrow into the soil, and resume growth whenever the pool fills again—a lifestyle called diapause. Snails will burrow into the sediment and secrete a mucus membrane across the door to their home, sealing in humidity (Colburn, Weeks, and Reed 2007).

When wandering dry summer woods, how would you know they once held pockets of water filled with any manner of creature? The shallow depressions remain, and much of the material that accumulated during the wet period will oxidize during the dry period, creating a noticeably lower spot. Also, the water may have left a mark—distinct bands found at the same height on a group of trees. Trees can signal waterlogging with trunks that grow wide at ground level and sport very shallow roots. You may see wetland plants growing in these "dry" spots. If you sift through the leaf litter in such places, you may find caddis fly larvae cases; if you dig in the soil, you may find dormant fingernail clams. Some of the depressions may fill again in fall, the "autumnal" phase of a vernal pool (P. Zedler, 2003). In these depressions you might be surprised to find a marbled salamander (*Ambystoma opacum*) keeping her clutch of eggs moist with her body and blending in with the bottom of the basin rather than advertising her presence with garish spots. These salamanders, found in southern New England and west to Michigan, mate and lay eggs in the leaf litter of dry pools. The eggs hatch when the pool fills in the fall and the larvae overwinter under the ice (Paton and Crouch 2002), gaining a head start on their spotted cousins in the spring. But it's not just amphibians that depend on these shallow woodland pools — these pools are also important for some very rare reptiles.

Looking for Turtles in Big Puddles

Author Sharon writes: Given the scrum of photographers around the hapless spotted turtle, one might have mistakenly thought we'd stumbled across a celebrity bathing in the shallow pond. Catherine and I had joined Jenn Jones to survey turtle traps in the swamps and vernal pools of Townsend, Massachusetts, and had tallied numerous painted turtles and a few snapping turtles; but it was only the spotted turtle that received the red-carpet treatment when we ultimately found her.

Our excursion's first stop of the day began in a maple swamp along a tributary of the Squannacook River (see chap. 5 for a discussion of red maple swamps). This piece of Massachusetts is quite unlike the location of Scott's spadefoot toad habitat, a cleared agricultural opening surrounded by upland forest. Here, the Squannacook River meanders among subdivisions as it travels between Townsend and the Nashua River. Bordering state and local forest preserves allow the river to flood and sustain the swamp. Further from the river, pockmarking the upland forests, are numerous small kettle holes and vernal pools.

Copious rainfall the day before meant our knee boots were woefully inadequate for traversing unseen logs and branches, but we pressed on, slowly making our way past cinnamon fern (*Osmunda cinnamomea*) and purple iris (*Iris prismatica*). We followed Jenn, aka the turtle whisperer, who was conducting a turtle survey for the Massachusetts Department of Fisheries and Wildlife, the primary interest being Blanding's (*Emydoidea blandingii*) and spotted (*Clemmys guttata*) turtles. Jenn has a reputation for unfailingly locating turtles, but that reputation was compromised by recent rains and deep waters; despite the cans of anchovies she left as bait, the circular-net turtle traps came up empty. At last, after wading through a few hundred feet of thigh-high water in a newly flooded beaver pond, we found, adjacent to a large, fallen tree, one submerged trap laden with painted turtles (*Chrysemys picta*). Painted turtles may be common in number and therefore not an "exciting" find, but against the deep browns and greens of the swamp, the wetted yellow-and-orange markings of the turtles were brilliant. Their yellow-striped heads tucked into red-orange-trimmed shells and their feet swam madly as we pulled them from the netting. Males were distinguished from females by their relatively long toenails, and then admired and tallied before we released them to disappear into water the color of vanilla extract.

Feeling somewhat more accomplished, we managed to pick our way out of the swamp without tripping or breaking the "don't get your underwear wet" rule. Once our knee boots were emptied, we headed for the vernal pools scattered throughout this complex of red maple swamp and dry-pine-and-oak woods. We walked up and over small, formerly glaciated ridges, into small depressions among red maples (*Acer rubrum*), American beech (*Fagus grandifolia*), and shrubby meadowsweet (*Spiraea alba*). The depressions were filled with spring rains that make islands of moss-covered boulders, isolated trees, and fallen limbs. Such places are often overlooked as insignificant, but here were the snappers. As fun as it was to find snapping turtles (*Chelydra serpentina*) in the traps, it was no picnic getting them out. For their part, the turtles were absolutely affronted by their capture and turned to face us with mouths poised to take any finger or toe that came near. A particularly large and outraged snapper refused to take instruction on how to exit the webbed-net trap. Despite the heavy, lethargic-looking body adorned with great folds of skin at its neck and an alligator-like tail, that hooked beak was fast—and its neck was longer than you'd think. Fortunately Jenn just needed to make note of its presence without bothering it to take measurements, and we moved out of sight while she tempted the turtle to go after a stick, leading it out of the net.

After the snapper's release, we moved over and down a hill across dry, sandy ground blanketed in pine needles to a near perfectly round depression in a one-acre clearing among red maple trees and meadowsweet shrubs. The water was at least waist-deep and smooth as a mirror, unbroken by floating or standing vegetation. There were no turtles, but we waded the edge admiring the lush lake sedges (*Carex lacustris*) and marsh ferns (*Thelypteris palustris*). The contours of this pool were shaped by a chunk of ice left behind by the last retreating glacier, forming the kettle hole now brimming with water. We proceeded down another steep incline toward a power line clearing and found another, smaller basin, about seventy-five feet in diameter, this time filled with bulrushes (*Scirpus* sp.) and ringed by highbush blueberry (*Vaccinium corymbosum*) and the ubiquitous meadowsweet. Jenn waded into the thigh-deep water to haul out the baited net-and-wire contraption. Here we found our A-list celebrity of the day, temporarily trapped in her breeding pool, the turtle sporting an orangey-yellow-polka-dotted black shell. She was the size of an adult hand, with spots on her head and orange-and-black legs; and she was rather cooperative, given her situation and the paparazzi juggling for a picture.

Jenn took the turtle's measurements and marked her shell before placing her on the soft sphagnum that borders the pool. Incredibly, it took only seconds for her to burrow into the sphagnum and disappear — we could not confidently determine just where she went, so we removed ourselves from the bank of the pool very slowly and very carefully. Stars do tend to get special consideration, but the spotted turtle's celebrity status is unwelcome — resulting from declining water quality, capture for the pet trade, and as always, loss of habitat. In this case the term *celebrity* is code for "of special concern," "threatened," or "endangered," depending on which state the turtle calls home. In Connecticut, Frank Golet's backyard spotted turtles (the sun turtles mentioned in the introduction) are tagged as a species of special concern.

Some Cannot Live by Water Alone—Vernal Pools Are Necessary but Not Sufficient Habitat

"Tested to withstand pressure from diving birds," says the sales pitch for the tiny GPS units scientists attach to frogs and other small creatures to track their movements. As if life as an amphibian were not tough enough. Of course, you can mark turtles the old-fashioned way, with colored markers or shell notches, as Jenn Jones did; but Robert Baldwin and colleagues (Baldwin, Calhoun, and deMaynadier 2006) wanted to track wood frogs, and you can't write on a wood frog. So these very patient scientists caught forty-three frogs and put tiny little belts on the less-than-three-inch-long frogs to track them via radio telemetry. They found that wood frogs in southern Maine migrate from their vernal pools to upland habitat and back again an average 385 meters, or the length of three and a half football fields! On a distance-per-gram basis, this is about the same distance as the annual round-trip that caribou make moving from their breeding to feeding grounds. Another study (Madison 1997) reported emigrating spotted salamanders moving an average of 118 meters (eight meters past the back goal line, or 387 feet) to their overwintering grounds.

Compiling a number of studies examining travel distances of amphibians, two researchers in Missouri (Rittenhouse and Semlitsch 2007) found that 95% of frogs and salamanders inhabit an area up to 664 meters from their starting pool. That's about the length of six football fields! Frogs, it turns out, are the amphibious creatures that step out the farthest, going twice the distance from home as compared with the salamanders. And the reptiles? A spotted turtle will travel a total distance of up to 1,680 meters (just over a mile) during one season, while the Blanding's turtle wins for most traveled, with 6,760 meters (4.2 miles) (Joyal, McCollough, and Hunter 2001). These last two figures mark the distance these turtles travel from wetland to wetland and from wetland to upland during part of the year in which they are active.

What these intrepid wildlife trackers are telling us is that seasonal pool residents travel. They may travel overland to the next seasonal pool or permanent water, pausing in the upland forest to estivate (go briefly dormant during the summer), or they may travel to forested uplands and wetlands to hibernate over winter. Scientists have found (Steen and Gibbs 2004) that there is a male survival bias in populations of snapping and spotted turtles in areas dense with roads; the higher death rate among female turtles might be a consequence of females traveling farther distances to find suitable nesting sites.

Now, think about this and pace these distances when you encounter a vernal pool, or measure the distance on an aerial photo or satellite image. How far can a frog, salamander, or turtle go before encountering a road, a clearing, a plowed field, a lawn, a parking lot? Roads can be barriers to travel, potentially preventing mating and egg laying, or a death trap—most assuredly preventing mating and egg laying.

To reduce the annual roadway slaughter, Brett Amy Thelen of the Harris Center for Conservation Education in New Hampshire organizes the annual Salamander Crossing Brigades, operations aimed at reducing road mortality for spring migration time. On warm rainy nights in the early part of spring, volunteers help move salamanders and any other amphibians across roads—in 2018, brigades throughout the region moved nearly eight thousand amphibians. For many volunteers it is their first time seeing a spotted salamander, and it's a "huge aha moment," Brett says. "It's something they remember forever." She tells the story of a couple with two young children who, while driving through an area where the salamander brigade was working (there are signs warning drivers), came to a stop in the middle of the road right in front of a spotted salamander. "The dad was so excited, he took the kids out of the car to look at the salamander—he had never seen anything like it, and he'd lived in the area for twenty years." The family pulled the car over, and Brett spent time talking about the salamanders and why the brigade was out on the road. "The whole family had their minds blown by this animal, and they'll never forget that."

What is necessary is an interconnected network of upland and wetland habitat for seasonal pool–dependent species to complete their life cycle, much like the landscape we traversed through near Townsend, Massachusetts. Maybe it's better to think of vernal pool organisms as landscape matrix–dependent species, and what we need to protect is a network of migration-connected habitat elements. To put it more plainly, we need to protect habitat diversity on our landscape. Vernal pools require us to think broadly about habitat and at different scales. Vernal pools are essential, but not sufficient for the survival of many of the species that live, breed, and feed in them.

Wetlands That Aren't Always There

There is such phenomenal diversity for what many might consider a big puddle. But the same temporary nature that makes vernal pools unique habitat is also their downfall. Add that they are often small and isolated and you have a recipe for anonymity or insignificance in the eyes of those who would use that supposedly empty space for something else. As with the woodland vernal pools' ephemeral cousins, the vernal pools of California and the playa lakes of western Kansas, Oklahoma, and Texas, some people don't believe in these sorts of wetlands—the wetlands that aren't always there.

The ephemeral wetlands of the drier Great Plains and in the Central Valley and coastal plateaus of California are also subtle gems scattered throughout their respective landscapes—landscapes dominated by grasses and cropland rather than trees. California pools are mostly small in size, like their woodland counterparts in the north woods, but fill with water during winter rains and remain wet rather than frozen through the winter before drying out during the hot summer. The playas of the southern High Plains can trend much larger, averaging 6.3 hectares (16 acres) (Guthery and Bryant 1982), and may fill with water only once every few to several years. What all these wetland systems share is the temporary, shallow nature of their hydrology, the unique repository of flora and fauna they cradle, and the threat of disappearing altogether.

In California and much of the West Coast, the rains start in November with sudden deluges bucketing down. Crossing the rolling hills of the Central Valley at night during such a rainstorm, visitors may be startled to encounter masses of migrating California newts (*Taricha torosa*) weaving their way through the grasses as they head to their breeding pools. At Jepson Prairie, in Sacramento Valley, the "temporary pool" at its center is large enough to deserve a name, Olcott Lake. After winter rains fill the lake to its brim, allowing tiger salamanders and fairy shrimp to breed, the seasonal cycle begins, California-style. Drying starts in March, and as the waters recede, the lake is ringed with a succession of flowers: first the whites of tiny popcorn flowers (*Plagiobothrys undulatus*) and meadowfoam (*Limnanthes douglasii* subsp. *rosea*), then the sunny yellow of goldfields (*Lasthenia* spp.), and at last the periwinkle-petaled *Downingia* spp. By June, the water is gone and the low pool blends in with the rolling dry prairie, golden hills dotted with clumps of oak. Once grazed by tule elk and set afire by native people, this vernal pool/ wet prairie escaped the damage of the plow that has converted most of California's prairies into valuable farmland. Jepson Prairie reminds us that this agricultural landscape hides precious wet pockets of biodiversity.

On a summer drive in the High Plains from western Kansas down to the panhandle of Texas, you are forgiven if you think there can't possibly be wetlands amid all the corn, sorghum, wheat, and cotton fields. Out in the pastures there are numerous farm ponds, but these are generally formed from dammed ephemeral creeks. While technically classified as wetlands by the US Fish and Wildlife Service National Wetlands Inventory, farm ponds rarely serve as functioning wetlands. But what the crops hide during the summer is very apparent during the spring rains, especially from an aerial view. Fly a plane across this region after enough spring rain and the number of shallow, rain-filled basins is absolutely stunning. This is what migrating waterfowl and shorebirds see: a bonanza of stopping places to rest and feed on their journey north. What the birds instinctively know is that, when the rain comes, the invertebrates emerge. What lays hidden in the clay-colored, cracked, and empty basin or field of corn in the summer is a myriad of invertebrates who have weathered dry spells in a dormant form (aestivation) or as eggs—they form a kind of animal seed bank (Boulton 1989).

The playas are also breeding refugia for amphibians, much like the vernal pools. Here you'll find the plains spadefoot toad (*Spea bombifrons*), the Great Plains toad (*Anaxyrus cognatus*), the plains leopard frog (*Lithobates blairi*), and the barred tiger salamander (*Ambystoma tigrinum mavortium*). Being that these creatures are located in an area of the country dominated by agriculture and few people, there is relatively little danger of them being squashed under car tires; but they are just as sensitive to changes in hydroperiod, the length of time the pool is flooded. And out on the plains, eroding sediment from the surrounding cropland is filling up the shallow basins, thereby shortening the flooding period and reducing the time available for pool inhabitants to breed, lay eggs, and hatch (Venne et al. 2012).

In California, vernal pools have been overrun by agriculture, especially in the Central Valley, but also by urbanization. Landscape changes lead to erosion, changes in water flow, and exposure to pesticides, as well as invasion by nonnative species, and continue to threaten what little remains of this distinctive ecosystem. Between 1994 and 2005, approximately twenty-five thousand acres of vernal pool habitat was lost across California as a result of residential, commercial, and industrial development projects. Depending on the region of California, estimates of vernal pool loss range from 75% in the Central Valley to 100% in areas of southern California; twenty plant and animal species associated with vernal pools are federally listed as threatened or endangered (US Fish and Wildlife Service 2005). Many species of plants and animals are endemic—found nowhere else but in these pools. Almost all the plants in the vernal pools are annuals, some of whose seeds can remain in the soil seed bank for decades, awaiting the right conditions to grow and bloom, hidden from view all the while.

Back in New Hampshire, Brett Amy Thelen says, "It's like a treasure hunt." In addition to salamander-saving brigades, she organizes and trains citizen scientists to inventory vernal pools; because "you can't protect them if you don't know where they are." She notes, "If you are not looking for them at the right time, then it is very easy to miss them. And they are not easily identified via remote sensing, so you really have to get on the ground to look for them." Workshops teach ecology and egg mass identification, and focus on getting people on the ground in their communities looking for vernal pools and documenting their location. It's important to document them in the right season, because activities that threaten vernal pools take place year round, often when it is difficult to detect the pools. Over the past ten years, trainees have identified two hundred pools in the Monadnock region of southwestern New Hampshire.

People who attend the training are often just interested members of the public, but some participants are members of their town conservation commission who want to learn more and then go back to their towns to document pools. Brett remembers two retired sisters who attended the training looking for something to do in their retirement that was out-of-doors. The very day after training, one of the sisters found a vernal pool, and Brett remembers her response: "I see them everywhere I look now. I just never knew what I was looking at — I never knew they were places of importance." That was four or five years ago; the sisters have been back every spring to volunteer for the inventory project. Brett relays another story, of a landowner who came to the training thinking he might have a vernal pool on his land but who just did not really know what a vernal pool was. After the training, he had a closer look at his property and realized that he did indeed have a vernal pool, finding wood frog and spotted salamander egg masses. Every previous year he had "tidied it up," removing all fallen woody debris — debris that serves as scaffolding for salamander and frog egg masses. Now he leaves the woody infrastructure and tends the vernal pool as critical habitat rather than treat it as a spring-cleaning chore.

In the town of Groton, Connecticut, Robert Ashworth's neighbor dumps her leaves and Christmas trees in the "empty" hollow adjacent to her house that sits on a forested lot. She has done this year after year, essentially burying the hollow under layers of debris that were too thick, and then remarked to Robert one spring that she no longer sees frogs and salamanders on her property. Ignorance of wetland existence and of wetland values is still quite common, according to Robert, who has served on Groton's Inland Wetlands Regulatory Agency for more than fifteen years (and is author Sharon's father). Asked what he knew about wetlands before serving on the agency board, he responds, "I had no idea what a wetland was, and I never thought about them being critical. I now will fight to save any piece of property just to have the wetlands remain intact. They're absolutely essential." Robert initially volunteered for the agency to generically "save the environment," and since joining has taken state-sponsored wetland classes. He regularly interacts with scientists regarding applications to the agency to fill wetlands. "People [property owners] see these little pink flags outlining the wetland, but then remark that there is no water. Yeah, but wait till spring and there will be water. People want to see water, but once they don't, they think it's free game to do whatever they want to do."

Robert explains that homeowners are generally cooperative when their projects come before the agency, but some fight any restrictions on property use. "They see this open space behind [their house] that they can put their shed in or pool, regardless of their deed that says it's a restricted area." Also, people want to go in and clean out the "messy place" by clearing out logs and fallen trees or mowing the rank grassy area to "make it look nice," or getting the muck out and putting "good solid dirt" in its place. Robert sardonically notes that once the "good dirt" is in place, it is planted with grass and sprayed with pesticides. Of course, discovery of these unpermitted activities merits a fine and orders to restore damaged wetlands — often at great cost.

The town of Groton is running out of buildable space and larger developers fight for every inch. When major developments come before the wetlands agency, the most common outcome is a reconfigured project, following lots of rancorous, late night meetings with scientists and lawyers on each side. All the agency can do is keep them out of the wetlands. Like many townships, Groton has no mandatory buffer area (protected area) around regulated vernal pools but can establish one as part of the permit requirements. Such latitude is given to many local conservation commissions throughout the Northeast although some townships have adopted vernal pool setbacks. The state of Vermont protects a buffer area of at least 50 feet around a vernal pool; Maine extends its protective barrier to 250 feet. Buffer areas are at least some measure of protection from stormwater laden with chemicals and sediment, but are little solace to the creatures that depend on the surrounding uplands as much as the ephemeral pool tucked among the trees. In New Hampshire where Brett lives and works, there is no statewide protection or mandatory buffer area around vernal pools, but the township can adopt protections. One of Brett's Conservation Commission trainees returned to her town and documented six new vernal pools. She is now advocating for a town ordinance to protect vernal pools.

Brett thinks the reason people get so excited about vernal pools is that they are an "approachable" ecosystem—they are small with relatively few critical species to learn. "Also," Brett points out, "it's spring, and everybody in New England, by the time April comes around, is so ready to be outside when there is no ice or snow. And so they get so jazzed about the spring amphibian migration and the follow-up vernal pool piece, because they've just been waiting for months and months to not be cold."

"However," Brett warns volunteers, "once the black flies come out, it's not fun anymore," so it's best not to procrastinate.

CHAPTER 7

Salt Marshes: A Disappearing Act

Most of the early settlements along Connecticut and Rhode Island
were on the coast because of the salt marshes. They were used by all the
farmers or anyone who had a cow or two, or 30 or 40 cows, or beefers,
or any kind of livestock. If they didn't have this salt marsh, it was a dead
issue. You just could not survive without it. There was no way.

—JOHN WHITMAN DAVIS, the last salt hay farmer in Connecticut,
"The Wisdom of 'Whit' Davis"

Imagine floating with the incoming summer tide. First, you creep up on the mudflats, the water filling countless burrows housing softshell clams, segmented marine worms, and amethyst gem clams. You are suspended over periwinkles, moon snails, blue mussels, sand shrimp, and spider crabs. Joining you are the fish: flounder, scup, dogfish, and sea robins. Still floating inland, you reach the low marsh, dominated by smooth cordgrass (*Spartina alterniflora*) that tolerates the twice-daily onslaught of salt and flood. A sturdy grass, reaching heights up to six and a half feet, it withstands pounding waves and steadfastly holds the soil with its roots, all while deprived of oxygen (see chap. 1, box 2). You and the tide infiltrate the marsh, filling up creeks populated by mummichogs and stickleback fish, eels, and smelt. The tide tops the creek banks and drops sediment among the cordgrass stems. On a normal day, you and the tide slip back out to sea, taking with you some of the detritus of the marsh. The nitrogen and phosphorous contained in this detritus will feed the base of the ocean's food web: the phytoplankton. Following your retreat, the shorebirds—plovers, sandpipers, dunlins, and willets—probe the mud and sand for its tasty residents.

Riding the tide back into the marsh at the new or full moon (when moon, earth, and sun are in alignment), you find yourself reaching further landward, flooding an area of the marsh dominated by saltmarsh hay (*Spartina patens*), inland salt grass (*Distichlis spicata*), and a short form of smooth cordgrass; this is the high marsh. The plants here can't tolerate the daily flooding of the low marsh. Buoyed by higher soil oxygen levels, these finer-leaved grasses can outcompete the taller, sturdier *S. alterniflora*. Looking out over extensive areas of high marsh, you get the impression of strong, sudden down bursts of wind that sweep the grasses in all directions. A common name for saltmarsh hay is cowlick salt hay, and it does appear as though giant cows roamed the fields licking sections of grass, plastering them this way and that; but it is just the mark of the swirling tides. Within the dense grass, if you look closely, you'll see marsh snails, spiders, horse mussels, and periwinkles. Listening closely, you'll hear redwing blackbirds, song sparrows, meadowlarks, and, if you're lucky, a clapper rail. Turn your eyes skyward to watch the swallows winging low across the marsh feasting on insects and the marsh hawk hunting for a meadow mouse or shrew.

As the tide recedes to the sea, some ocean remains behind in pannes or pools in low spots where the only escape for water is evaporation. Only the most salt-tolerant of plants survive here—glasswort (*Salicornia depressa*), seaside arrowgrass (*Triglochin maritima*), sea blite (*Suaeda* sp.), and the seaweed knotted wrack (*Ascophyllum nodosum*) (see box 4). You remain also, wishing you could flow in and out of this lovely marsh every day, riding through another thousand years of tides. But times are changing. On the edge of the marsh are tupelo trees, pitch pines, and bayberry shrubs, if there is any vegetation at all. More often you spy concrete walls, backyards, an airport, even an industrial park. These precious shorelines are crowded now, the marshes long ago filled for buildable land, seawalls, and dikes built

Box 4. Adaptations to Salt

Common names for the glassworts (*Salicornia* spp.) include *sea pickles* and *sea beans*. While the monikers *pickles* and *beans* are botanically misleading, the plants are good eating and taste of the sea. Various cookery websites recommend the succulent, salty plants cooked, raw, or pickled, and as especially good on salads. There are four species of *Salicornia* in New England, and they all look like slender, green to red, segmented fingers sticking 4 to 20 inches (10 to 50 cm) up out of the sandy salt pannes of the high marsh.

Glassworts and other plants that tolerate high salt levels are called halophytes. Such plants adapt to high salinities by excluding salt from roots and leaves, excreting salt, storing extra water, or transporting and storing salt. For example, *Spartina alterniflora* accumulates salts in its cells to draw in pure water from seawater. The concentration of salts in the plant's cells is higher (and thus water molecules are fewer) than that found in seawater, causing water to flow from the seawater into the plant cells. (This movement of water from high concentration to low concentration is called osmosis). Plants that do not tolerate salty water will wilt because the water in their cells moves out into the seawater, causing the plant to lose turgor (firmness caused by water pressure in the cells). Plants like the glassworts instead excrete highly saline water through their leaves; when the water evaporates, the salt crystals are left behind.

to block incoming waters—all to disastrous effect for salt marshes up and down the coast.

Nuisance Arising from Swampy Lands: The Pine Creek Marsh Story

The title of this story comes from an 1895 Connecticut colonial law; the statute is still on the books.

> **Connecticut General Statutes, chapter 368e, § 19a-212 (2013):**
> **Nuisance arising from swampy lands**
>
> When there exist upon any premises swampy or wet places or depressions in which a foul and unhealthy condition, arising from natural causes, permanently exists, the director of health of the town or the health commit-

tee, director of health or board of health of any city or borough, in which
such places or depressions exist, upon the written complaint of any person
and upon finding that such places or depressions are a source of danger to
the public health, may cause such places or depressions to be filled with
suitable material or drained.

As with most salt marshes up and down the northeastern coast of the
United States, the story of Pine Creek Marsh in Fairfield, Connecticut, is
a love-me/love-me-not tale, tracing the history of human attitudes toward
these coastal conundrums from the arrival of the colonists.

Prior to European settlement, the coastal areas of New England were
seasonal hotspots for native tribes harvesting fish, mollusks, crustaceans,
and waterfowl. Pine Creek Marsh was no different until European settlers
arrived, emigrating from Massachusetts to the great bounty of coastal Connecticut. The settlers wrested control of the area, then known as Uncoway,
from the Pequots in 1637 in a local battle called the Great Swamp Fight. The
Pequots vanquished, the émigrés garnered land through a treaty with the
local Pequannock tribe and established what was to become the town of
Fairfield in 1639.

> The waters brought forth abundantly "various kinds of fish—shad in
> prodigious quantities, but bass were the fish they caught most plentifully,
> taking in at Black Rock sixty or eighty in a night; occasionally some of
> them weighing as heavy as twenty-eight pounds. Clams, oysters and escal-
> lops more than could be eaten." Eels and smelt swarmed in the waters.
> Whitefish were so plentiful that they were drawn in by nets, and distrib-
> uted for manure upon the lands. Beside these, lobsters, crabs, mussels
> and other inferior shellfish were found in great quantities. The fresh
> water streams afforded trout, lamper-eels and turtles of considerable
> size. (Schenck 1889, taken from an early resident's journal)

The coastal marshes furnished resources beyond what settlers could
catch for dinner. Marsh peat provided fuel (Schenck 1889), and the grasses,
saltmarsh hay (*Spartina patens*), salt grass, and blackgrass (*Juncus gerardii*)
provided fodder for grazing sheep and cattle. The settlers both hayed the
marshes and turned their livestock out on the flat grassy areas (Rozsa 1995).

Long, narrow ditches were dug soon after settlement. This allowed water to flow out more easily, thus lowering surface water levels, providing easier access for haying equipment and promoting the growth of saltmarsh hay (Rozsa 1995). As useful as the marshes were, a portion of the Pine Creek Marsh was inevitably described as "waste-meadow." In the mid-1600s, the first dam was proposed to block incoming tides (Schenck 1889) and create a freshwater meadow to replace the saltmarsh hay (Steinke 1988).

Mosquito vs. Human: Early Perspectives on Coastal Salt Marshes

Mosquitoes have always been a part of salt marsh ecosystems, but a concerted effort to eradicate the pest began when soldiers returning to Connecticut from the Civil War brought with them malaria. Malaria was known to occur in Connecticut for 250 years prior to 1900, but with the soldiers' return, the disease reached epidemic levels (Wallis 1960). By 1900 the tiny, annoying insect had been confirmed as the malarial carrier, and the filling and draining of wetlands began in earnest. Despite the freshwater origins of the disease-carrying culprit—the common malaria mosquito (*Anopheles quadrimaculatus*)—all flooded areas, whether fresh or salty, were targets for mosquito control. A 1915 state law authorized cooperation with towns and cities for control of mosquitoes, and an all-out effort left most of Connecticut's salt marshes ditched. Most of the ditching was done by hand in the 1930s, by men in Depression-era work programs. If ditching was not deemed sufficient mosquito control, dikes—walls built to keep the tide at bay—with tide gates were built to further dewater the marsh. "Before such control, it has been said that one couldn't tell the color of a cow in the pasture near the shore until after the mosquitoes had been brushed away" (Wallis 1960). For good measure, state experts recommended that oil be dumped into the ditches to control the "wigglers" (mosquito larvae). Such activities initially reduced populations of both the freshwater common malaria mosquito and the saltwater pest species *Aedes sollicitans*.

People continued to pour into these coastal communities. Marshes were filled for development or dredged for harbors; those left were ditched and sprayed for mosquitoes, then most were diked and gated to block the tides. Fifty percent of tidal marshes between Southport Connecticut and the Connecticut River were ditched by 1900 (Rozsa 1995).

The first flood-protection gates at Pine Creek in Fairfield were installed in 1938, preventing the tide's access to approximately 25 acres of a 124-acre coastal marshland. Behind the flood-protection gates, "new land" was now available for houses, parks, ball diamonds, dumps, and the Fairfield Lumber Company. Having initially built out onto the undiked salt marsh on fill, the lumber company's stock would float away on an occasional storm tide; so in 1958 the town built a set of flood-control gates downstream, adding another thirty-seven acres of buildable land (Thomas Steinke, pers. comm., May 13, 2015; Roman, Niering, and Warren 1984).

As more people piled onto the coast and built permanent homes to replace summer homes, the remaining marshland behind the tide gates began to change. The sea no longer reached into the diked marsh and rain flushed the marsh soils of salt. Fresh water from surrounding upland drained into the marsh and was held there. Saltwater fish, mollusks, and crustaceans disappeared. The winged fauna, the waterfowl, herons, and shorebirds that fed on saltwater creatures disappeared. The vegetation underwent a drastic transformation. No longer supporting saltmarsh hay, the drained and diked marsh filled with an aggressive type of common reed (*Phragmites australis*), an invasive stowaway on ships arriving with European settlers (Saltonstall 2002). Rather than the fine, two- to seven-foot-high grasses that rippled in the sea breeze, there was now an impenetrable twelve- to fifteen-foot-high wall of thick-stemmed reed (see chap. 1 for more about common reed). Come winter, that reed became a fifteen-foot-high standing mass of dry fuel; and ten years after building the floodgates, the lumber company burned to the ground in a phragmites fire (Steinke, pers. comm.).

After all efforts at eradication, the mosquitoes at Pine Creek were worse than ever. Garbage and winter road sand filled up mosquito ditches no longer scoured twice daily by the tides. The fish that dined on the ditch-dwelling mosquito larvae vanished. The detritus of reed grass clogged not only the ditches but also the pipes meant to pass water from the increasingly impervious upland to the sound. Stormwater running off the ever-increasing acres of pavement and rooftops caused back flooding of storm drains, resulting in pressure in stormwater pipes great enough to blow the manhole covers off the streets in heavy rainstorms (Steinke, pers. comm.). The phragmites grew so thick that mechanical ditch maintenance became too costly and time-consuming. DDT was substituted for mosquito-eating fish and tides.

And then there were the fires. Dense stands of phragmites, dead and dry in the winter, were perfect tinder for almost yearly conflagrations.

> *Phragmites* fires are short-lived but exciting. A strong March wind will blow a wildfire across a 20 acre marsh in 20 minutes — faster than you can run through the 12 ft stems, with each stem 3 to 8 in apart, with the leaves slicing away at your hands and ears, and as the dry peat duff collapses underfoot and your lungs fill with smoke and dust. . . . Over the years in Fairfield, these fires have burned a local lumber company, consumed out-buildings, cars, porches and fences, scorched homes, cracked window glass, and melted the vinyl siding from houses. (Steinke 1988)

Fill, Fill, Fill: "Solving" the Flood and Bug Problem

What is a town built on a salt marsh to do? Dig out the marsh peat, dig out underlying sand and gravel to use for road base, and then fill in the holes with town garbage. Better yet, also address the stormwater flooding by laying down five-foot-diameter pipes in a trench from the coastal road to the sound, and then fill in over the pipes with garbage. Fairfield built numerous, smaller flood-relief dikes that were "strategically located across the marshes and creek channels so as to provide additional marsh reclamation for sand and gravel, garbage disposal, marina, golf course, park and single-family home development" (Steinke 1988).

Still there was flooding. The more development and "improvements" achieved, the more the flood and erosion control board would accommodate demands for dikes around newly built properties. Each new dike increased the height of tidewater on remaining unprotected properties closer to the sound. Normal high tides now flooded these properties. In addition to increasing the height of the high tide and its associated flooding in the un-diked areas, the diking also reduced the volume of water entering and draining from the salt marshes during a tidal cycle (called the tidal prism), and that resulted in less flushing and scouring of the creek channel downstream, resulting in the channel slowly filling in with sediment and causing boats to run aground. Many boat owners with keel sailboats changed over to motor-boats due to the shoaling of the channel after diking (Steinke, pers. comm.). And the cycle repeated, creeping ever closer to the shore: dike, fill, make the marshes "useful" again. However, this business as usual cycle was about to come to an abrupt halt.

Massachusetts was the first state to adopt legislation establishing conservation commissions, giving municipalities the power to acquire land for passive use rather than for traditional ballparks and playgrounds. Connecticut followed suit, and Fairfield was one of first towns to set up a conservation commission, in 1966. At this time, one of Fairfield's first selectmen, John Sullivan, believed the town needed to save some of the undeveloped land appreciated by the town's residents, which would also attract new residents (Steinke, pers. comm.). He appointed like-minded people to the first Conservation Commission; with new federal and state funding, they put a plan together linking open spaces with stream corridors, lakes, and marshes. In a few short years, Fairfield had accumulated six hundred acres of upland and marshland. Of course, now the town needed a land management plan. With the Yale School of Forestry just up the road in New Haven, two young graduate students, Whitney Beals and Peter Westover, were contracted to develop a management plan. The result was *The Pine Creek and Mill River Watersheds, Fairfield, Connecticut: An Ecological Guide to Open Space Land Use*, or the "red book" for short, published in 1971. Salt marsh restoration was the plan's centerpiece.

At the same time that Beals and Westover were writing recommendations for preserving and restoring salt marsh, the flood and erosion control board was promoting further "flood-control" efforts through the time-honored method of building dikes and tide-blocking gates. The Conservation Commission weighed in with desires for a floodgate that would allow tidewater into the marsh during normal tides, but would block incoming water when it reached a certain height. The flood-control engineers came back with: Well, sure we can do that, if the floodgates are controlled by gas-powered generators built on reinforced-concrete dikes and maintained by a crew with salaries and benefits. And so, in 1968, a once-and-for-all dike was built across the main channel of Pine Creek at the foot of the marsh, sealing off from the ocean all but ten acres. Fairfield's tally of lost salt marshes was now 61% of the original acreage (Rozsa 1995). The only nod to the Conservation Commission's desires was to modify the traditional tide gate with a hinged cover so that it could be opened and closed manually. Manual manipulation failed, and the gates stayed closed on the Pine Creek Marsh.

Ironically, 1968 was the same year that the state of Connecticut passed the Tidal Wetlands Act, which at least legally recognized the importance of these systems and required approval by the state to conduct any development activities in wetlands, including filling and dredging (Rozsa 1995). Up and down the coast, city boards charged with flood and erosion control were stopped in their tracks. As Tom Steinke, retired director of conservation for the town of Fairfield, describes it: the eastern version of the "Wild West" had come to an end. Now towns would have to buy fill for roads, send their garbage elsewhere, and find upland for future parks and ball fields — all more expensive than using the "wasted space" of those nuisance marshlands. Many towns, including Fairfield, continued to fill, ignoring state law for the next couple of years. But then came the inspectors and enforcement officers from the Department of Environmental Protection tallying violations and informing towns that they must dig up their illegal, putrefying garbage dumps and haul them away. Fairfield countered—what if we left the garbage and in compensation restored marsh elsewhere? The state agreed and almost before the cement on the new flood-control dike had cured, the city of Fairfield started making plans to restore tidal floodwaters to Pine Creek Marsh.

Olives to the Rescue: Letting the Tide Back In

This is the situation that Tom Steinke found himself in 1971, having just received his master's degree in wildlife ecology from the University of Massachusetts. He took the two-year job in the town's conservation department as filler between graduate school and a "real job." The Conservation Commission handed him the red book for a job description. Implementing the red book meant taking out the dike and floodgates built just four years previously. It was an untenable position; the town governance realized that the salt marshes were the most important natural feature in the area, meaning most problematic as well as most promising. But how to solve the integrated problems of phragmites fire, mosquitoes, stormwater and tidal flooding, habitat degradation, and illegal fill caused by flood interventions? And how to do this in the face of the tightly held belief that the only thing that stood between one's property and the force of the sea was the flood-control dike? Tom needed to find a way to let enough salt water onto the marsh each day to keep it healthy, while also allowing stormwater to flow out and preventing catastrophic storms from damaging inland properties just beyond the marsh.

Tom's father was a mechanical engineer, and Tom had spent his youth tagging along, reading engineering plans; thus he had developed a knack for solving technical problems. What the situation at hand needed was a two-way flood-control gate—fresh water out, tides in—but without the necessity of a crew and gas generators. The gates would have to open and close on their own, but how?

Olive barrels. Olive barrels were just the thing. Tom's father-in-law had an import business in olives, and those olives were shipped to the United States in large plastic barrels, like the ones used for homemade rain barrels. Tom strapped a couple of barrels to the hinged floodgate covers with rope so they would float on the tides and move the gate cover up and down. After two years of various barrel float configurations and failures, Tom had a working prototype. Town officials and engineers cast a skeptical eye at the jury-rigged, multicolored olive barrels roped to the tide gates. Nevertheless, Tom had a working model of a self-regulating tide gate in 1976 and installed it in nearby Ash Creek for a test run.

It worked beautifully. The gate had two floats (olive barrels); one would open the gate as the tidewater rose, but if higher storm tides came in, a second float would overcome the first and close the gate. The effect of returning salt water to the marsh behind the gate and dike was dramatic. Each successive growing season, the phragmites grew shorter and less dense, the salt-marsh hay returned, the tides scoured the mosquito ditches, the fires ceased, and the mollusks, crustaceans, and birds began to recover. However, while the marsh was taking care of itself, the hardest part of Tom's job now was convincing coastal communities to adopt the new tide gates.

Tom laughs when he talks about the commission meetings he was asked to leave. Who was this guy coming here telling us we needed to let the sea back into the marshes? Have we not solved the flooding problem with our dikes and gates? The mosquito problem belongs to the health department, the fires belong to the fire department, the storm sewer–flooding problems belong to the public works department. Tom spent his evenings at meetings: commission meetings, town meetings, neighborhood meetings, meetings in firehouses, libraries, and people's living rooms. He'd take them to look at the new self-regulating tide gate at Ash Creek, and they'd look at the olive barrels from Greece and Spain, and laugh. He'd demonstrate how they worked and seine the clean mosquito ditches to show how the aquatic life was recovering. He showed them a pipe gauge he'd installed to track the declining height of the phragmites, and point out the recovering saltmarsh hay. He'd tell them how the fires had ceased and that there were no reports of flooding.

Upstanding Suburban Vandals Get the Message: Understanding and Accepting Tides

Some people were convinced, others not, and so a covert battle ensued over the traditional tide gates still installed in the channels of the marshes. Those who wanted marsh restoration would go out to the conventional tide gate under cover of darkness and open the hinged gate completely, bolting it open with a chain. Of course, if the gate was left in this position when the next storm blew in, those living close to marsh were flooded. Residents who did not want marsh restoration (often those who lived closer to marsh) would go down to the gates under cover of darkness with bolt cutters to close the gates, and then drive big metal pipes down in front of the gate so that it could not open. Of course, if a big rainstorm came, then the water could not escape the marsh and still they were flooded. Having otherwise upstanding citizens resort to vandalism was driving city conservation commissioners crazy, because it was they who got all the phone calls. The commissioners, in turn, called Tom.

So it fell to Tom to manage the tide gates—but he was ready. On the heels of countless meetings and demonstrations, he convinced the city to install his new self-regulating tide gate on one of the smaller channels of Ash Creek. This new tide gate was enclosed in a locked chamber, preventing access to those with chains, bolt cutters, and pipes. To assuage the doubters, Tom cut an opening in the locked cover of the gate so that anyone could reach down, grab a rope, and pull the gate shut on their own. He told people that if they were nervous, they could come on down and pull the rope and the gate would close. Tom would visit the gate from time to time and sure enough, people would show up before a storm, reach in, and pull the rope. Then, one night, Tom was at the gate during a rising storm tide—but the nearby residents who typically showed up to close the gate were late. By the time they arrived, the gate had closed on its own, just as designed. Steinke was right! After eight years of effort and a patented new invention, Tom finally got permission to install the new tide gates in the channels of Pine Creek Marsh, and in 1980 the town of Fairfield breached the 1968 cross-channel dike.

Since that time, towns up and down the coast have installed Tom's self-regulating tide gates, and the recovery of salt marsh ecosystems has been nothing less than stunning. A study conducted in the first years following the breach of the 1968 Pine Creek dike found that, after just one year, phragmites density declined by 50%, and its height was reduced by two meters (6.5 feet) (Roman, Niering, and Warren 1984). Twelve upland plant species, mostly asters and goldenrods, disappeared from the marsh; natural salt marsh plants and animals recolonized the wetland.

> Incidental dip-net samples indicated that grass shrimp and mummichogs were some of the earliest animals moving into the marsh creeks followed by worms, amphipods and then mud snails, ribbed mussels and fiddler crabs. Various herons, egrets, resident and migratory geese and ducks, horseshoe crabs, snapper blues, black-backed flounder, and commercial quantities of shellfish have accompanied the restoration of the marshes as well. (Roman, Niering, and Warren 1984)

Tom's self-regulating tide gates have been adopted in coastal areas worldwide, although sadly they are no longer built with olive barrels. However, as salt marsh restoration efforts expanded throughout coastal New England and recovery in formally closed marshes accelerated, another bleak trend was beginning in the tidally influenced marshes that remained—they started dying.

Where Have All the Marshes Gone?
The Story of the New England Salt Marsh Die-Off

> Something is killing New England's salt marshes, and scientists are trying to figure out how large the problem is, and how to stop it. Parts of the marshes, normally teeming with cord grass, fish and birds have turned mud brown and bare of life except for fiddler crabs. . . . "We're talking

about a crime scene investigation, some forensic ecology, if you will,"
[said Ron Rozsa, coastal ecologist with Connecticut's Department of
Environmental Protection].
—ASSOCIATED PRESS, "Conn. Scientists Investigate Marsh Die-Off,"
June 26, 2006

It turns out, the tale of the disappearing New England marshes and the likely cause is a window into salt marsh ecology and the scientific process. The story begins before anyone really noticed. It would be twenty years before someone looked at and interpreted the satellite pictures, but pieces of the marshes had been disappearing since at least 1976 (Coverdale, Bertness, and Altieri 2013). In 2002, the phenomenon took on a sense of urgency when the National Park Service documented a 12% loss of salt marshes along Cape Cod National Seashore and scientists in Connecticut documented losses on the shores of Nantucket Sound (Smith 2006). On the ground, salt marsh death resembled a dark-brown bathtub ring riding the banks of salt marsh creeks and mosquito ditches. To the casual observer it might have appeared that water had receded, exposing bare mud. Speculation as to the cause included drought, sea level rise, rising water temperatures, ice damage, fungal infection, and eutrophication (Smith 2006; Lewis 2007; Alber et al. 2008). In 2007, the *Boston Globe* (Lewis 2007) interviewed a scientist at Brown University in Rhode Island about another interesting hypothesis: the crabs did it. Dr. Mark Bertness postulated that the native purple crab (*Sesarma reticulatum*) was eating the dominant salt marsh cordgrass, *Spartina alterniflora*, effectively denuding creek banks of vegetation. Others were skeptical (a pathogen had been implicated in southern salt marsh die-offs), and there was no experimental evidence to support any one hypothesis.

Mark Bertness is a transplant from the Pacific Northwest; when he moved to Providence to take the job at Brown University in 1980, he found Narragansett Bay depressing. Compared to Puget Sound, Washington, the bay is a relatively recent ecosystem; having been scoured out by the last glaciation, it is full of "weedy things." "And," Mark continues, "you have the insult of the Industrial Revolution on top of it [with all] the pollution." However, the bay's comparative advantage is its simplicity, and here he could study the ecosystem engineering skills of the introduced marine snail *Littorina littorea*, the common periwinkle. The periwinkle, which originated in

Europe, now inhabits New England rocky shores; more than inhabiting the shores, Mark found, the snails maintained the shores, and without them the marshes filled with sediment. The idea that a consumer in the shoreline food web could significantly impact the structure of this system was rather heretical at the time. Of course, to get to the rocky shores of New England, one must pass by and through the salt marshes, a system Mark had been ignoring in the time he was studying the snail. But accumulated field trips meant years of casual observations, and then revelation. On the leading edges of the salt marshes are concentrations of ribbed mussels that appear to hold together the edge of the marsh. Back from the shore a bit was the fiddler crab zone. It looked as if these two animals significantly structured the marsh, which brought into question the notion that the salt marshes were ecosystems driven by physical processes such as nutrient input.

Traditional theory holds that salt marshes are controlled by what is called bottom-up management—its structure and function directed by such physical (abiotic) factors as nutrient levels, temperature, and salinity (Teal 1962). Consumers—the top feeders in the marsh food web (e.g., crabs, insects, and fish)—were considered less of an influence on salt marsh processes. But Bertness and his student, Brian Silliman, working in the southeastern United States, demonstrated that plant productivity there was instead governed by grazing snails and their predators (Silliman and Bertness 2002); more snails meant less plants. Bertness and others working in the Northeast (Bertness, Crain, et al. 2008) then published work supporting the hypothesis that increased nutrient input into coastal marshes (carried in by runoff from fertilized lawns and croplands) resulted in increased insect damage to vegetation. Plant hoppers and grasshoppers dined on the nutrient-enriched cordgrass and saltmarsh hay, reducing its growth up to 76% within two years. The researchers further suggested that the whole concept of domineering physical controls ought to be revisited in light of the potential for human disturbances (such as fertilizer runoff) to alter the impact of consumers on salt marsh ecology. By turning the bottom-up management theory for salt marshes on its head, Bertness and others (see Duffy 2002) laid the groundwork for testing the possibility of crab-induced marsh die-off.

Ecology Crime Unit: The Science behind the Puzzle

In the summer of 2007, the Bertness Lab at Brown University set out to test the purple crab hypothesis on Cape Cod, where crab densities appeared to be high, and along Narragansett Bay, where the extent of marsh damage was lower (see box 5). By surveying and comparing the two areas, this initial study (Holdredge, Bertness, and Altieri 2009) established a correlation between high crab densities and loss of cordgrass. They set up cages surrounding the cordgrass and excluding crabs, and observed that the cordgrass grew thick and lush, while the cordgrass exposed to crabs was completely devoured. So a high density of crabs did seem to coincide with marsh die-off. But what was driving the high populations of purple marsh crab? They suspected crab populations were controlled by predators—lack of predators, in the case of large crab populations—and set out to test the hypothesis by using captive crabs as bait. Picture a dog leashed to a stake in someone's yard. In the salt marsh version of this scenario, crabs were tethered to a stake by fishing line and allowed to roam an area that included their burrows. At each creek bank analyzed, researchers put a pair of crabs on tethers—one surrounded by a predator exclusion cage, the other not—and left them overnight. It doesn't take much imagination to guess what happened to the uncaged crabs; predation was noted in the lab book as "dismembered body" or "remnant leg and carapace." Comparing Cape Cod and Narragansett Bay, the researchers concluded that predation pressure on the purple marsh crab was much less on Cape Cod, where cordgrass die-off rates were highest.

So now we have the beginnings of a possible explanation for marsh die-off in northeastern salt marshes: fewer crab predators, which leads to more crabs munching on cordgrass, and eventually ends in a denuded marsh creek bank—a domino effect of consequences through the food web. Preliminary studies in hand, the Bertness Lab continued to investigate by replicating the predator exclusion experiments and further suggesting that intense recreational fishing of crab predators such as striped bass and cod was to blame for the reduced crab predator population, thereby triggering the ecological cascade ending in marsh die-off (Altieri, Bertness, Coverdale, Herrmann, et al. 2012).

Using fourteen sites on Cape Cod experiencing varying degrees of die-off, a crew of graduate students measured and compared predator numbers,

Box 5. A Foray into the Winter Marsh; or, A Lesson in Character Building

The first of Mark Bertness's students to work on salt marsh die-off was Christine Hol-dredge,* an undergraduate honors student in his lab. She was working in the Bertness Lab when Stephen Smith, plant ecologist at the Cape Cod National Seashore, asked Mark to come a take a look at what was happening to the marshes on the cape. After that eye-opening visit, Christine began investigating the crabs' relationship to marsh loss (see Holdredge et al. 2008). At the end of the field season, in early November, she and Mark went out to the town of Wellfleet on the cape. It was much colder and windier than back in Providence. They were underdressed but just had a couple things to pick up out of the marsh, maybe take a last measurement, given that they had driven at least two hours to get there.

Out on the marsh it was so cold and windy that it was difficult to move, so they decided to split up the tasks to finish more efficiently. They set off in different directions. When Mark returned to where Christine was to have finished her tasks, he discovered her twenty feet out in the marsh, stuck. To get the task done quickly, Christine had decided to do what they all did during the hot summers: cut across a shallow body of water, rather than go the longer way around. Of course in the summer, when they got hopelessly stuck in boot-sucking muck, they just got down on their hands and knees and sort of swam out of their predicament. On a cold November day, this tactic was not an option. Despite Christine's recent history as a pre-Olympic hockey team player, she had gotten about twenty feet across and had gotten stuck, and was now so cold that she could not move. Mark could not go in after her without also getting stuck—and the tide was coming in. Fortunately, as researchers do in remote study sites, they had a stash of supplies in the woods at the marsh edge: PVC pipe and some flat wooden boards. Mark was able to grab a couple of boards to lay across the muck, walk out, and reach Christine with a ten-foot-long piece of pipe that she used to pull herself out while Mark held the other end. "We had been so familiar with that site, but summer and winter are so different." She could have died.

To emphasize the physical rigors of salt marsh research, Mark relays a comment made by one of his graduate students (also a superb athlete, having served as a goalie for the Irish women's soccer team). Driving back from the cape during her first summer of research, and after a twenty-hour day, she said, "I thought this summer was going to be fun, but I did not realize I was going to find out who I am."

*Christine Holdredge Angelini, now an assistant professor of ecological engineering at the University of Florida.

rates of crab predation, crab densities, the quantity of cordgrass eaten by crabs, and the extent of marsh die-off. Investigating any potential link to recreational fishing, they spent what must have been some fine days of field-work counting the number of anglers out enjoying the marshes in July. Comparing historic to contemporary aerial photos, the researchers were able to estimate changes in recreational fishing by noting the change in fishing infrastructure (docks and boat slips) over time.

The conclusions reached by the research team describe a classic trophic cascade: marshes hosting recreational fishing had lower predator densities, high crab densities, higher rates of cordgrass herbivory, and a high incidence of die-off; marshes with no recreational fishing had higher numbers of predators, lower crab densities, and lower rates of cordgrass herbivory, and were intact.

The evidence for crab-induced marsh die-off was mounting, but a marsh eaten up by crabs would still have live rhizomes (underground stems) that could recover the following season; this was not the case in New England. So could the lack of post eat-out recovery mean that the crabs were not ulti-mately responsible? Carrying the research further, the Bertness Lab inves-tigated the extent and impact of crab herbivory not just aboveground but belowground. They indeed found significant feasting going on underground in the crab burrows, killing even the larger cordgrass plants normally big enough to escape aboveground dinner parties. Lacking soil-binding root masses and riddled with crab burrows, the salt marsh peat had been sub-stantially weakened, becoming soft and loose. Loose peat slumps from the creek banks, hampering any chance of a recovery the following growing sea-son (Coverdale, Altieri, and Bertness 2012).

By now it appears there is a solid link between recreational fishing and marsh die-off via high population densities of the purple marsh crab. But might this trophic cascade be unique to Cape Cod? Could evidence of the human impact on marsh die-off be generalized to other areas of New En-gland? Setting out again, the Bertness Lab members took the time-tested methods of past field experiments, surveys, and historical analyses and ap-plied them again to Cape Cod, but then expanded the research to include marshes along the shores of Narraganset Bay and Long Island Sound. The researchers concluded:

In southern New England, die-off sites had greater fishing pressure and more fishing infrastructure than vegetated sites. These conditions led to localized depletion of top predators including blue crabs, striped bass (*Morone sazatilis*), and American eels (*Anguila rostrata*). This predator depletion has had cascading effects in southern New England, where predation on *Sesarma* (purple crabs) decreased and *Sesarma* densities increased at heavily fished sites, and has led to significantly elevated grazing and the creation of large die-off patches. (Coverdale, Bertness, and Altieri 2013)

Reconstructing the trophic patterns of marsh die-off for Cape Cod and southern New England revealed that die-off began after construction of docks and marinas and intensified as coastal development increased. This historical pattern began over thirty-five years ago on Cape Cod and more recently in southern New England, with about a twenty-year lag time paralleling the increase in the number of marinas and docks. Unfortunately, this may mean that, without intervening management, southern New England is likely to experience continued loss of marshes.

A Reappearing Act: The Possibility of Salt Marsh Recovery

Curious thing about research—the more you know about a system, the more questions continue to arise. As the Bertness Lab focused on the disappearing marshes, they noticed something during the 2010 field season on Cape Cod: some of the dying marshes had begun to recover. The muddy peat that had slumped off the creek banks was being recolonized by cordgrass. Looking over aerial photos of marsh sites that had never had a documented die-off, as well as the sites found to be recovering, a pattern emerged. The recovering sites were clearly demarcated by three zones: the crabless, cordgrass-covered creek banks closest to the water; the denuded zone, loaded with cordgrass-eating crabs and burrows; and a high-marsh zone that was actively being eaten but still vegetated.

Back to counting crabs. The team found that the low-marsh zone near the water was not sturdy enough to sustain crab burrows and was avoided. Once cordgrass is able to obtain a toehold in the crab-free low marsh, changes slowly accrue as cordgrass alters the substrate for the better—capturing sediment and organic matter that enhance and strengthen the peat for further cordgrass expansion.

> However, the full recovery of New England salt marshes is not assured, nor will it be immediate. Although we observed a net decline in die-off areas, grazing continued in the high zone. Moreover, die-off on Cape Cod led to the loss of 150–250 years' worth of accreted marsh peat in many marshes (T. Coverdale, unpublished data). Since recovery areas in some marshes are small relative to the large areas lost to calving, slumping, and creek and ditch widening associated with human-triggered die-off, the full recovery of these marshes to their original extent will likely take centuries. (Altieri, Bertness, Coverdale, Axelman, et al. 2013)

Now for the next, most recent twist. Recent marsh recovery on Cape Cod may be linked to the invasion of a nonnative species, the European green crab (*Carcinus maenas*). This is likely to be the next chapter in the New England marsh die-off saga. While the green crab was introduced to North America in the early 1800s, it is not common in healthy marshes. They are quite common, however, in marsh die-off areas riddled with purple crab burrows. While green crab densities did correlate with marsh recovery, a direct impact of green crab on purple crab needed demonstration. The researchers once again enclosed crabs in cages—purple crab only and purple crab with green crab. Forcing crabs to be roommates resulted, at a minimum, in purple crab evictions by the invading green crab. Eviction was not the only consequence, as indicated by severed purple crab legs strewn about the shared domiciles. It appears that the green crab has reinstated predation pressure on the purple crab and "is well suited to accelerate the recovery of heavily degraded salt marsh ecosystems in New England" (Bertness and Coverdale 2013). Finally, an invasive species that does some good!

Salt Marshes and the Rising Sea

> To stand at the edge of the sea, to sense the ebb and flow of the tides, to feel the breath of a mist moving over a great salt marsh, to watch the

flight of shore birds that have swept up and down the surf lines of the
continents for untold thousands of years, to see the running of the old
eels and the young shad to the sea, is to have knowledge of things
that are as nearly eternal as any earthly life can be.
—RACHEL CARSON, *Under the Sea Wind*

It's become the clarion call of coastal ecosystem management: coastal systems protect shorelines and buffer storms! Hurricanes Katrina and Sandy renewed public attention and interest in coastal wetlands. With coastal communities feeling a bit more vulnerable, more funding came through for marsh study and restoration. Pair this vulnerability with future climate change scenarios that include more intense storms and sea level rise and you've got more than a few recipes for disaster. But you've also got a heady mix of research and problem solving focused on the shores of our continents.

Overall, wetland vegetation on the coast absorbs tidal surges: plants slow water velocity and reduce water turbulence, exerting a drag on passing waves (Gedan, Altieri, and Bertness 2011; Pinsky, Guannel, and Arkema 2013). This effect does vary, however, and depends on the surrounding landscape and a storm's intensity, size, and track (Wamsley et al. 2010; Pinsky, Guannel, and Arkema 2013). Belowground, the roots of marsh plants do bind the soil, resisting loss of land to the sea; but wetland vegetation has its limits in the face of strong oncoming storms. Recommendations to bolster marshes include setting aside larger parcels and pairing the natural protective properties of coastal wetlands with more traditional hard structures like seawalls. Such plans always look fabulous when artfully illustrated or as fancy 3-D models, but large-scale projects remain intangible as we dither about costs.

The field of ecological economics can interpret scientific findings and put them in a dollars-and-cents context that policy makers understand. A team of economists, ecologists, and geographers found the average annual value of wetlands in the state of New York to be just over $51,000 *per hectare, per year* (Costanza et al. 2008). If you total up the value of coastal wetlands in the entire Northeast, it's $105,333 per hectare (approximately 2.5 acres, or about the size of two football fields) per year.

But what happens to the shore if marshes disappear?

The Sparrow That Does Not Sing: Salt Marsh as Critical Habitat

It's a mild and sunny summer day on the tidal salt marshes at Barn Island Wildlife Management Area, which sits across Little Narragansett Bay from Stonington, Connecticut. Walking through a wide, dry stretch of marsh, Chris Elphick, a conservation biologist at the University of Connecticut, focuses a spotting scope on a group of little brown birds hidden among the thigh-high grasses. Chris identifies the rare saltmarsh sparrow by the yellow shading on its face and the crisp dark streaks on its breast. Saltmarsh sparrows—which represent the most vulnerable of many species that call this habitat home—make their nests in the high marsh among stems of saltmarsh hay, escaping the twice-daily tides that flood the lower marsh. But the Barn Island high marsh and others like it are disappearing as the rising ocean brings salty tides farther inland. Chris gives the rare sparrow thirty to forty years before it disappears from the planet.

The best chance this little bird has of avoiding a flooded nest is to lay eggs immediately after the high spring tides, with most birds laying after the new moon flood tides. Successful parents will have built the nest high enough in the grasses to escape flooding, but low enough so as not to be seen by aerial predators. Birds that miscalculate will re-nest in hopes of fledging a new batch of youngsters before the next high tide. The consequences of unexpected events are dire for such small, rare populations, and in 2009 and 2010 Connecticut marshes experienced higher-than-usual tides. In 2010 Chris and his students located two hundred nests but only five fledglings.

The band of habitat these birds occupy is shrinking as sea level rises and the high-marsh zone disappears. For four thousand years, coastal marshes have kept pace with gradually rising oceans by retreating inland or growing vertically. In this latter process, called accretion, sediment drops out of slowing river waters as they reach the sea, building new substrate for marsh plant colonization and adding to the existing marsh surface when water overtops creek banks. The sediment that makes it out further in the ocean, and that which is sloughed off the seaward edge of the marsh, is washed back with the tide and deposited on the surface of the marshes, falling out of the water column and trapped by vegetation. Add deposited sediment to the buildup of organic matter and the surface of the marsh increases vertically, raising the elevation of the marsh.

But the world has changed: the rate of sea level rise has doubled on the northeastern US coastline since 1990, dams keep fresh sediment loads from the coast, and human structures such as roads and seawalls block inland migration. Consequently, as the sea creeps ever landward, permanently flooding the lower marsh, the dominant lower marsh plant *S. alterniflora* sends its rhizomes landward, displacing *S. patens* and the saltmarsh sparrow's habitat.

Drowning in Place—Nowhere for the Marshes to Go

Sea levels have been rising since the last ice age peaked more than twenty thousand years ago — as massive amounts of ice melted into the sea and then as water has warmed over the centuries. Balanced by continental rebound, coastal salt marshes made a brief appearance until they were overwhelmed as the sea overtook the land. Peat from these older salt marshes can now be found sixty-one meters (two hundred feet) underwater off of Cape Cod (Warren 2014). Then, about four thousand years ago, sea level rise in New England slowed to one millimeter a year, and the salt marshes we used, then destroyed, and now desperately want to save, began to form. The rate of sea level rise had eked up to about 2.6 mm (0.1 inches) a year by the time we started measuring in the 1930s. Marsh accretion rates, however, were able to keep pace. Then, in the 1980s, the rates of sea level rise along the mid to north Atlantic coast accelerated to just over four millimeters (0.16 inches) a year; many scientists speculate that this is just too fast.

Those who seek to protect coastal ecosystems look for ways to reduce barriers to marsh migration, finding spaces for marshes to move inland instead of drowning. One promising approach involves cooperation between conservationists and city and state planners. Using a coastal mapping tool originally developed for the US Environmental Protection Agency called the Sea Level Affecting Marshes Model, or SLAMM, federal and state agencies and conservation organizations create maps showing where high tide will be as sea level rise increases. Having that information allows communities not only to plan for future urban infrastructure, but also to make room within that infrastructure for salt marsh habitat.

Rhode Island has used SLAMM to project ocean flooding for all twenty-one of its coastal communities. "We found it was a place to start the sea level rise conversation," says Caitlin Chaffee, a policy analyst with the Rhode

Island Coastal Resources Management Council (CRMC). "[It pinpoints] opportunities to remove aging infrastructure and to accommodate the wetlands."

The city of Warwick, for example, worked with CRMC to identify a number of crumbling roads that occasionally flood but will eventually dead-end in the ocean. To close portions of the roads and restore wetland habitat, however, the city needed local residents' buy-in, and the SLAMM maps proved convincing. "We also used the model to show that this particular area along the coastline in Warwick Cove will be inundated in five to ten years," says Warwick planning director William DePasquale. Given that reality, he says, it was relatively easy to convince people that it would be better for the city to buy and preserve the land rather than allow new development that would soon be threatened.

Marc Carullo, environmental analyst for the Massachusetts Office of Coastal Zone Management, says some communities in his area are excited about the model's potential to help them proactively plan for coastal changes. Currently, salt marshes protect homes along the coast by decreasing water speed and turbulence and diffusing incoming waves. "If we have large expanses of salt marsh becoming tidal flat, we're going to lose ecosystem services and that could play a big role in how exposed that homeowner is to storm surge," Marc says. The visualizations SLAMM provides, he says, will help motivate communities to protect salt marshes and the services they provide.

Back in Connecticut, SLAMM modeling shows a 50–97% loss of high marsh by 2100. That's up to ten thousand acres. The model predicts the loss will be mitigated by less than a thousand acres of potential wetland "gain" at higher elevation — and Chris Elphick says even that is optimistic, since he sees little evidence of marsh migration today except for salt marsh grasses colonizing coastal lawns. He and the little brown saltmarsh sparrow he seeks to protect—indeed, everyone and everything that benefits from the services salt marshes provide — can only hope that the awareness the modeling brings will help coastal communities find a place for salt marshes as they accommodate sea level rise.

Further down the Atlantic coast, the city of Wilmington, North Carolina, is expected to experience "one-in-one-hundred-year" floods in thirty of the fifty years between 2050 and 2100, to coincide with what seems like a modest prediction in average yearly rise in sea level, 0.42 to 1.32 centimeters (0.16 to 0.52 inches) (Kopp et al. 2015). Not only are extreme flooding events more frequently moving onto higher ground and flowing over larger areas, but the occurrence of "nuisance" flooding events is increasing (Sweet et al. 2014). More-common minor flood events in developed areas overwhelm stormwater management capacity, damage roads, and lead to deteriorating infrastructure. Salt water that cannot move inland and flood marshland will collide with our built environment.

Moving off the Atlantic coast and into the Gulf of Mexico, there are locations where humans have not built up to the dunes or in the salt marshes. Do the salt marshes move inland where they have room to breathe? Yes, depending of course on topography and fresh water flowing from the uplands. Along the low-sloping coastline of what is called the "Big Bend" of Florida's west coast, salt marshes are disappearing at the water's edge as the sea rises, but moving inland and replacing coastal forests (Raabe and Stumpf 2016). Overall this has meant a net gain in salt marsh, as expansion into coastal forests has exceeded marsh loss to open water. This is not the case on the Gulf coast near the Mississippi delta, where lack of sediment supply has starved marshes of their ability to accrete and match the rising seas.

Draining and Filling Salt Marshes to Save Them

Fly over or meander through any salt marsh along the Atlantic coast and it is hard to miss the miles of parallel ditches that mark the wetlands. The parallel scratches in the surface of the marshes are mosquito ditches, dug first in the 1930s to drain the marshes but now to create watery habitat for the mummichogs and other fish that eat mosquito larvae wriggling around in the water. Along the eastern coast of Florida and in the Gulf coast, impoundments became the preferred method for mosquito control when ditches proved unsuccessful (Rey et al. 2012). Many of the ditches are part of the open marsh water management (OMWM) system that was adopted in the late 1960s along parts of the northeastern coastline to reduce populations of disease-carrying mosquitoes. Over the years, marsh management methods have been tweaked and refined as they spread through the northeastern and mid-Atlantic states. The techniques involved in OMWM are now the predominant methods of mosquito control along the eastern coast and employ selective ditching and ponding rather than wholesale attempts to drain water off the marsh, as had been done starting in the 1930s. Understanding the life cycle of one of the most problematic mosquito species, the saltmarsh mosquito *Aedes sollicitans*, provides the rationale for today's version of mosquito control: *A. sollicitans* does not lay its eggs in water, but rather in mud.

The saltmarsh mosquito breeds in the high marshes; low marshes are flushed twice daily and are not good mosquito habitat. In the high marsh, female mosquitoes lay their eggs in mud that will remain dry for several days. When the tides of the new and full moons reach the high marsh, the larvae hatch, grow, pupate, and then become mating adults over the course of just a few days. A mated female then flies out of the marsh in search of blood to nurture her eggs (which sounds like a reasonable maternal instinct unless that blood meal is you). The mosquitoes have adapted to the wet-dry cycles of the high marsh. The trick with OMWM is to reduce the area of open mud and bring in predators such as the aptly named mosquito fish (*Gambusia* sp.).

The concept is not entirely new, but as practiced in the 1950s and '60s, mosquito control involved holding water in salt marsh habitat with the added bonus of producing waterfowl and fish habitat. These impoundments have now been removed in favor of restored tidal influence, which leads us back to prime mosquito habitat. The compromise now is to create small pools in the high marsh—just deep enough not to dry out and expose mud for egg-laying mosquitoes. The pools are often connected to existing creeks so that the tide fills them up with water as well as mosquito-eating fish and invertebrate mosquito predators. Additionally, pools are created by plugging old mosquito ditches (Lesser 2007). However, ditch plugging may cause excessive waterlogging of the marsh surface in some locations (Adamowicz and Roman 2002). One of our current marsh problems, as discussed above, is flooding of the high marsh. So while on one hand agencies responsible for mosquito control are creating pools, agencies responsible for marsh restoration are attempting to drain pools. In most cases, draining means excavating small "runnels" from the flooded pool in the high marsh to existing ditches or creeks to facilitate drainage. The goal is to restore vegetation to the formerly flooded area and once again promote marsh accretion.

But, of course, there may not be time to "promote" marsh accretion, so scientists are proposing to combat sea level rise with a wholesale effort to save a marsh by replacing lost sediment with dredged material to raise large expanses of marshes above the encroaching sea. Using techniques developed in the Gulf coast, fine-sediment slurry is sprayed under high pressure onto the surface of wetlands that are sinking below rising seas. After years of regulatory enforcement of no-fill-in-the-wetland rules, we are now doing just that to save them. Climate change makes more than the weather weird.

From Greenhouse Gases to Blue Carbon

Salt marshes, as well as mangroves and seagrass beds, don't just play victim in our changing climate—they have a major role to play in carbon sequestration.

Blue carbon is the term given to the carbon sequestered and stored in marine and coastal environments (Nellemann et al. 2009). Plants and algae obtain carbon in the form of carbon dioxide (CO_2) and incorporate it into the structure of the plant or alga. This is carbon sequestration. The plant uses some of the carbon for energy, and, as its roots respire, CO_2 is released into the soil. When the plant dies, the carbon in its tissues is consumed by the detritivores and decomposers in the soil, which release the carbon into the soil as CO_2. The CO_2 from root respiration and decomposer respiration then passes from soil pores into the atmosphere. This is carbon emission. When plants are in an anaerobic environment, such as that found in wetlands, organisms requiring oxygen cannot decompose the plant, so most (but not all) of that carbon remains in the soil. This is carbon storage.

One hears a great deal about the tropical rainforests as great carbon-sucking ecosystems, but the world's primary ecosystems for carbon sequestration and storage are our coastal salt marshes, mangroves forests, and seagrass beds (Mcleod et al. 2011). Salt marshes in particular sequester five to eighty-seven teragrams of carbon per year—that's *tera*grams (one teragram is equal to one million metric tons). For comparison, gasoline burned for transportation in the United States produced about 11 teragrams of CO_2 in 2016 (US Energy Information Administration 2017). Salt marsh plants have been sequestering carbon and then dying for hundreds of years, building peat to keep up with sea level rise. So now, our modern-day salt marshes are sitting on a major storehouse of carbon, up to six meters (approximately twenty feet) thick in some places (Chmura 2013). The value of salt marsh carbon sequestration and storage is just now being recognized. While the science is still new, blue carbon is figuring into calculations of how much carbon sequestration is worth in terms of cold, hard cash. The monetary value of carbon sequestration (Sutton-Grier et al. 2014) may soon be considered an additional factor in climate change accounting (Ullman, Bilbao-Bastida, and Grimsditch 2013).

Unfortunately, all the excitement over coastal systems finally getting their due recognition in the carbon cycle is diminished by the realization that we are losing these carbon sponges. Worse yet, their destruction could result in increasing carbon emissions. Disturbances expose surface peats and sediments to oxygen, allowing oxygen-loving, decomposing bacteria and fungi to feast on and release all that carefully stored carbon back into the atmosphere (Pendleton et al. 2012). Destruction of a salt marsh either directly (filling) or indirectly (death by crab) obliterates carbon sequestration as the plants die off and a rapidly rising sea can mean the end of marsh accretion or the possibility of inland migration.

Lest we end our tour of wetland systems on such a sour note, we begin the next chapter in a salt marsh off the coast of Maryland where inspiration, curiosity, and science blend, giving rise to the field of wetland restoration.

CHAPTER 8

Wetland Restoration: Changing Techniques, Changing Goals, Changing Climate

> This is a story of the creation of a tidal marsh by three
> Ph.D. chemists who had never grown a plant and, at
> the beginning, knew nothing about wetlands.
> —EDWARD GARBISCH, "Hambleton Island Restoration"

So recalls Edward Garbisch in his description of one of the first formal attempts in the United States at wetland restoration (see Garbisch 2005). Garbisch writes that, after reading the book *Life and Death of the Salt Marsh* by the husband-wife team of John and Mildred Teal (see Teal and Teal 1969), he packed up his professorship at the University of Minnesota and headed back to his boyhood home on the eastern shore of Maryland. It was 1971, and Garbisch set his sights on an eroding island off the coast of St. Michaels in Chesapeake Bay. Joined by his former students Paul Woller and Robert Mac-Callum, and bolstered with the necessary Army Corps of Engineers permits, Garbisch set out to demonstrate that what had been destroyed could be repaired. At the time, few restorations had been attempted, although in North Carolina William Woodhouse and his colleagues had begun experimenting with using salt marsh vegetation to stabilize the sands, gravels, and sediments dredged up in channel-building operations (Woodhouse, Seneca, and Broome 1972).

Garbisch includes some pictures from those days in his 2005 summary of the project, although *project* is a rather benign word for this early herculean effort. Hambleton Island had been breached and split into two islands, so the idea was to restore the protective tidal marsh and repair the breach. The images depict what many of those in the ecological sciences will recognize as the often grubby, arduous tasks of "fieldwork"—three young men in T-shirts and cutoffs transporting loads of gravel and sand by small barge and unloading them by hand and hose along the shores of the eroding island. *Five hundred* barges of sand and gravel. The sand and gravel was populated with sixty thousand individual wetland plants, plant by plant. Of course, at that time, in the early '70s, there were no nurseries from which to order plants for restorations, so the trio had to grow their own from locally collected seeds. It was the beginning of the first wholesale nursery for native plants.

By 1973, the sand-and-gravel fill was in place and anchored by row upon row of vigorously growing cordgrass, saltmarsh hay, cattail, common reed, and salt grass—the beginnings of a new tidal salt marsh in place of that which had been lost. Then the geese came. Again, as anyone familiar with fieldwork will understand, nature happens despite best-laid plans or carefully plotted research designs.

> In early April 1973, a flock of Canada geese (*Branta canadensis*) flew in to the wetland on a moonlit night during a high tide and fed on the marsh for several hours. As a result of this eat-out, the monitoring program and all of the planned research abruptly terminated. (Garbisch 2005)

Those attempting to restore lost ecosystems are nothing if not creative (and obstinate). While the research protocol had flown off with the geese, the event prompted various inventions to prevent a repeat eat-out, and by 1976 an aerial photograph shows a robustly vegetated salt marsh doing its job—providing marsh habitat and protecting the shoreline. Garbisch went on to found the not-for-profit wetland restoration and research center Environmental Concern Inc. Environmental Concern has since become a native plant nursery and has worked to restore over seven hundred acres of shoreline wetlands in and around Chesapeake Bay.

These early days of restoration were all about trial and error. Garbisch and his team did not have books on native plant propagation and planting; they had to make it up as they went along. Suzanne Pittenger-Slear, president and CEO of Environmental Concern, relates a story from those ad hoc days. Garbisch, looking for something to hold young salt marsh plants, grabbed his wife's just-purchased rubber doormat to use as a plug tray. As Suzanne states, "All of the resources that we have available to us now were not available to Ed, so actually using a doormat was probably not a bad idea." And the "invention" that prevented geese from eating future restoration efforts? Sticks and twine: two-inch-by-two-inch wooden stakes, placed twenty feet apart, strung with two lines of twine.

As for the Hambleton Island restoration site itself, Suzanne reports that it is still holding its own; but, sadly, most of the island shore continues to erode. Maintaining restorations is "not unlike a garden," Suzanne says. Environmental Concern is called back to monitor projects, but maintenance at the most basic level is necessary, like removing shoreline debris so new plants can grow. Fortunately, Suzanne reports, they are asked to return for maintenance and monitoring, and for the most part people understand the "garden" concept. As the miles of shoreline restored increase, so does the public's understanding and interest, and Suzanne is impressed by the growing interest in native plants. Education is a key component of the restoration process. Suzanne recalls one of her favorite projects, when Environmental Concern was asked to plant shoreline across the bay from St. Michaels in 2015—with 350 ninth graders from the Howard County schools. It was a project that involved the entire staff of Environmental Concern and included continuing education back in the classroom.

While the experiment of wetland restoration has moved on beyond doormats to ubiquitous plug trays, the legacy of that first attempt lives on in those soon-to-be-graduating kids, and in miles of shoreline revegetated and protected by sticks, twine, and heart.

The Early Days of Wetland Restoration

bogs of treachery, mires of despair

—J. LARSON and J. KUSLER, describing historic views of wetlands in *Wetland Functions and Values: The State of Our Understanding*

The first wave of European settlers to New England depended on the coastal salt marshes for fish and fowl harvest and hay for livestock. As described in the previous chapter, salt marshes and the inland wet meadows provided for livelihoods; before long, though, the burgeoning settler populations began to see wetlands as more bother than benefit. By 1645 Boston and other growing eastern cities were filling marshes for industrial expansion or dredging them for larger harbors (Vileisis 1997). Europeans fanned out over what would become the United States, draining and filling wetlands as they went. "Vile," "evil" swamps bogged down travel, sheltered horrible creatures, and bred disease. The prairie wetlands could not be planted and were prone to unsuspected floods. The Swamp Land Act of 1850 turned over federally owned swamps to the states explicitly for drainage (Beck 1994). Few voices cried out in protest, but by the late 1800s and early 1900s, some began to notice and despair at the wanton destruction of nature, including still reviled swamps. While the restoration of ecosystems would not gain traction for another forty to fifty years, landscape architects were experimenting with natural gardening, and from within the naturalized school of landscape design rose the beginnings of ecosystem restoration (Egan 1990; Jordan and Lubick 2011).

The first intentional attempt at restoring a native plant community may have been Frederick Law Olmsted's restoration of the Back Bay Fens in Boston (Egan 1990). Olmsted did not set out to restore a marsh but was commissioned to create a city park from the fetid, low-lying areas of the Back Bay neighborhood. The area was a former salt marsh cut off from tidal influences in 1821, partially filled for housing and serving as a garbage and sewage dump when Olmsted was brought in to deal with the problem in 1878 (Vileisis 1997). Although most people wanted a typical park, Olmsted sought to recreate the original salt marsh, which would also serve to hold back stormwater flowing in from Stony Brook (Egan 1990).

Foreshadowing the techniques pioneered in the Connecticut salt marshes nearly a hundred years later (see chap. 7), Olmsted and city engineers returned some tidal flushing to the marsh and reduced fresh water inputs by rerouting fresh water into the Charles River. A preeminent practitioner of the naturalized school of landscape design, Olmsted chose native salt marsh grasses to bring back the look and feel of a natural marsh. The idea was not popular—one can only imagine a salt marsh in the city center now.

It took Bostonians only fifteen years to reroute the fresh water back into the fens and fill large portions of the marsh for more typical gardens and parks.

While very few in the late 1800s and early 1900s recognized the now familiar ecosystem services provided by wetlands — flood control and water filtration, for example — even urban dwellers could appreciate the birds that inhabited wetlands. Unfortunately, this appreciation extended to sporting beautiful feathers, and even whole dead birds, on one's head. Plumage hunting and mass-market hunting of waterfowl devastated bird populations, and calls rang out to limit the slaughter. Local Audubon Societies, along with sportsman's clubs and women's groups, agitated for laws to protect birds, and in 1900 the Lacey Act prohibited interstate traffic in birds and animals killed in violation of state laws (Vileisis 1997). In 1903 Theodore Roosevelt established the first national wildlife refuge to protect waterfowl at Pelican Island off the east coast of Florida. Of course, these initial steps did not curtail the downward trend in waterfowl population numbers, and sport hunters continued to sound the alarm.

Matters got worse during the droughts of the early 1930s; money for refuges was tight during the Depression. In 1934, Jay Norwood "Ding" Darling, appointed to the Bureau of Biological Survey (now the US Fish and Wildlife Service) by president Franklin D. Roosevelt, created the Federal Duck Stamp program to generate funds to purchase and rehabilitate land for waterfowl conservation. With the help of the Works Progress Administration and the Civilian Conservation Corps, dikes and ditches brought water to drained and drought-stricken land. Understanding the connection between waterfowl populations and their breeding areas, a group of sportsmen founded Ducks Unlimited (DU) in 1937 and the following year undertook to protect and rehabilitate the wetlands of Canada's prairie pothole region. Drained wetlands across the Canadian prairies were outfitted with small dams to trap water and recreate shallow marshes. The sign that hung at the entrance of DU Canada's first wetland restoration said it all: Big Grass Marsh Duck Factory No. 1 (Historica Canada, n.d.).

Most of the concern about wetland conversion and declining waterfowl populations resulted in the purchase of wetlands for conservation; restoration played a minor role. When restoration did take place, the emphasis was on dams for ducks. A typical restoration involved digging a pond or installing a low dam to fill a depression and then adding a dollop of land in the center for waterfowl nesting. As wetland science expanded and amassed data, the field of wetland restoration became much more sophisticated, moving beyond these early restoration attempts, which were fondly or derisively coined "duck doughnuts."

Into the Modern Age: New Wetlands to Compensate for Those Lost

A year after Ed Garbisch returned to the Chesapeake Bay to try his hand at recreating a salt marsh, Congress passed a number of key amendments to the Federal Pollution Control Act of 1948 (FPCA). While the activities of Garbisch and company fell historically under the purview of the Army Corps of Engineers (COE) — who regulate dredge and fill activities in waters traversed by boat and barge traffic — the new laws added in 1972 formally gave the COE the additional responsibility of regulating dredge and fill activities to protect water quality under section 404 of the FPCA. Under the new law, the COE was to regulate dredging (removing material from wetlands) and filling (adding material) in all "navigable waters" — a designation that included tributaries of larger rivers, regardless of whether one could travel them by boat. It took a lawsuit in 1975, Natural Resources Defense Council v. Callaway, to establish that section 404 was meant to include all waters of the United States, including adjacent wetlands. In 1977 President Jimmy Carter signed into law a reauthorized and amended FPCA, now called the Clean Water Act, that directed protections for "waters of the United States," explicitly including wetlands.

Meanwhile, scientists and natural resource professionals gathered in Florida in 1974 for the First Annual Conference on Restoration of Coastal Vegetation. Robin Lewis, a professor of biology at Hillsborough Community College in Tampa, had been experimenting with mangrove restoration and sought to gather those working in this new realm of ecology (Jordan and Lubick 2011). Papers presented at this conference were variously titled "Salt Marsh Creation on Dredge Material and Natural Shores," "Florida Department of Natural Resources Efforts in Coastal Vegetation Restoration and Marine Habitat Construction," and "Possible Use of Spoil Material to Replace Lost Coastal Vegetation in Florida" (*Proceedings of the First Annual Conference on Restoration of Coastal Vegetation in Florida* 1974).

Restoration results were coming out of the southern coastal marshes, and other parts of the country started to take notice. By the summer of 1973, inspired by Woodhouse's and Garbisch's successes, highway engineers as far away as Maine were reexamining the feasibility of salt marsh relocation and restoration in cases where new roads had destroyed existing wetlands (Maine Department of Transportation and Reed & D'Andrea 1974).

A new concept was slowly evolving: Mitigation, meaning to decrease or offset the negative impacts of wetland destruction. By 1977, the concept of restoration as mitigation was being written into development plans and state natural resource policies. Edward LaRoe, in a paper titled "Mitigation: A Concept for Wetland Restoration," presented at the National Wetland Protection Symposium in Reston, Virginia, outlined a requirement newly adopted for Oregon's Land Use Program—specifically, its estuarine resources goals:

> When dredge or fill activities are permitted in inter-tidal or tidal marsh areas, their effects shall be mitigated by creation or restoration of another area of similar biological potential to ensure that the integrity of the estuarine ecosystem is maintained. (LaRoe 1978)

Restoration as mitigation soon became the main driver of wetland restoration activities. In 1978 the Council on Environmental Quality clarified National Environmental Policy Act (NEPA) regulations formalizing a mitigation sequence that had been casually debated and applied in COE permit decisions (Hough and Robertson 2009). The mitigation sequence became part of the Clean Water Act in 1980 and is now known as the section 404(b)(1) guidelines (box 6). By 1986, the COE was required to demand mitigation as part of projects that sought to fill, drain, dredge, or alter wetlands. In 1989, the first Bush administration declared the "no net loss" policy, firmly establishing the barter of restored, created, and enhanced wetlands for existing wetlands. Soon, almost any builder whose project would destroy a marsh, wet forest, or sedge meadow needed to find a way to make amends in some other swampy spot, or even by constructing a brand-new wetland nearby.

Box 6. Section 404(b)(1) Guidelines

"Mitigation" includes:

   (a) Avoiding the impact altogether by not taking a certain action or parts of an action.

   (b) Minimizing impacts by limiting the degree or magnitude of the action and its implementation.

   (c) Rectifying the impact by repairing, rehabilitating, or restoring the affected environment.

   (d) Reducing or eliminating the impact over time by preservation and maintenance operations during the life of the action.

   (e) Compensating for the impact by replacing or providing substitute resources or environments. (Council on Environmental Policy 1978)

Delineation: Finding "De-line" between the Wetland and the Upland

The Clean Water Act protects most, although not all, wetlands, and state and local laws in many parts of the United States give additional layers of protected status to wetlands. Legal protection, however, does not mean wholesale prohibition of any type of impact; it means, for the most part, that someone wishing to fill, drain, dredge, or alter a wetland for a development project needs to get a permit. For some kinds of projects and some kinds of wetlands, these permits are difficult to obtain and are judged on a case-by-case basis (individual permits; for example, a housing development that is slated to fill in a portion of a wetland). Projects such as roads or pipelines, or construction that is considered to be beneficial to a large number of people or only minimally harmful to the wetland, are more likely to be permitted (general permits; for example, a gas pipeline crossing a creek). And in many permitted cases, mitigation will be required.

But, in any case, it is essential to know where the legal (formally, "jurisdictional") wetland begins and ends. Finding this legal boundary is called wetland delineation, which is as much art as science. To be considered a wetland, the area must have three important features: hydric (wet) soils, wetland plants, and wetland hydrology. Professional wetland scientists must seek specific pieces of evidence to prove whether the area in question has the soils, plants, and water characteristic of a wetland. Because a lot of money rides on the answer, extensive research and debate has gone into the complex guidelines that define these three characteristics.

Looking at a watery marsh full of cattails or stepping down a steep incline from a dry forest into deep water, the difference between upland and wetland is obvious. But in many other places, it is much harder to see the difference, and what looks like a perfectly dry clearing in the summer woods may in fact fill with water in the spring and fall (see chap. 6).

Beth Markhart, a wetland scientist in Minnesota, describes one such situation, when she was flagging the wetlands within a proposed utility corridor along the Little Amnicon River, a tributary to Lake Superior in Wisconsin. The landscape here is complex, she explains, with "really subtle patterns and mosaics of different kinds of wetland communities, interfaced with upland prairies, agriculture and pastureland. In this area, the capillaries of the landscape come together in the forested headwaters, this is where they start to channelize and flow into more permanent tributaries." Black ash swamp blankets much of the area, grading into a wet aspen forest on the dry side and alder swamps on the wet end. Like many consultants, Beth was drawn into wetlands work when the Clean Water Act regulations were passed, and has lived through the evolution of wetland regulations, learning and applying the three-part criteria for delineating wetlands and then passing that knowledge on to new trainees. The area along the Little Amnicon River was the first foray into delineation for one young man—a rough place to start. "We spent an entire week documenting finely detailed upland-wetland boundaries, mapping wetland communities and assessing habitat values," Beth says. "It was fun to help my young colleague, fresh out of college, develop this awareness of how many things you have to think about—the interaction of the topography and the soils and the vegetation. Not many field exercises in colleges these days are very skill-oriented, so he was quite overwhelmed by having to identify and quantify such subtle changes in the landscape and map them using the GPS. It was shock and awe for him. He kept thinking he was never going to get it—I told him it wasn't always this difficult! After a few days, he began to see the subtle patterns, and it felt like victory for both of us."

This area around Lake Superior has a red clay soil substrate, which hides all the key features needed to tell upland soils from wetland soils. In this location as in some others, the determining factor between upland and wetland is often just the water-holding capacity of the soil; there may be no change in the tree canopy at all. The only changes in the vegetation are inconspicuous shifts in the understory layer: herbs like large-leaved aster (*Eurybia macrophylla*) inhabit the uplands, while sedges, monkey flower (*Mimulus ringens*), marsh hedge nettle (*Stachys palustris*), and others that can tolerate wet conditions are found in the wetlands.

Fortunately, not every wetland is so difficult to delineate. Tom Peragallo, a soil scientist from New Hampshire, describes the normal process for finding the wetland boundary: "I typically start on the wetland end of the transition zone, and then I walk perpendicular from the wetland to the upland, probing with a soil auger until I see where the soil changes from hydric to nonhydric," based on color, texture, and depth. If the vegetation in that spot is no longer wetland vegetation, he hangs a flag to indicate the wetland edge. Every plant species in the United States has been categorized by its level of affinity for wetlands, ranging from obligate wetland plants—found in wetlands 99% of the time—to upland plants, found in wetlands less than 1% of the time. Wetlands must have over half their dominant plants on the official wetland plant list, which is maintained by the US Fish and Wildlife Service.

Providing evidence of wetland hydrology is another one of the challenges for wetland delineators. Tom notes that, "in problem situations—particularly, former wetlands—we have to look at historic aerial photography and do more work to document the wetland hydrology as well as the soils. Where areas have been filled or altered, sometimes we have to bring in a backhoe to dig a soil pit to reconstruct where the former wetland boundary was."

Wetland Soils

Probably the most overlooked component of wetlands is the soil, often hidden below murky brown waters and lush vegetation. In the wettest situations, the lack of oxygen (see chap. 1, box 2) slows down the process of decomposition, so bits of dead plants and other debris drop down and don't decay very quickly. This results in thick spongy layers of organic peat. Where there is a little more oxygen or other elements to assist in decomposition, the soil will consist of rich layers of dark brown–black muck, often underlain by grayish clay, silt, or sand. *Muck* is actually a technical term, for highly organic, well-decomposed soils. Deep layers of black muck or peat are classic indicators of wetland conditions, as many young wetland scientists quickly learn.

"Way back when I was just beginning as a professional soil scientist, we were out characterizing some organic soils. I was told to watch out for 'black-leg disease,'" Tom Peragallo says. Having no idea what his colleagues were talking about, Tom just shrugged and followed his mentors. "We did some sampling, digging holes to examine the wetland soil, then backfilling the hole with the soupy organic soil." As they were gathering their equipment and cleaning up, Tom accidentally stepped right into the newly filled hole and sunk in, up to his waist, smearing his trouser legs black with muck. "So I learned the hard way about black-leg disease, and they all got a good laugh," Tom recalls.

These black and dark-brown colors, and the washed-out grays below them, are the hues of really wet soils. Upland soils tend to be more colorful—reds and yellows from oxidized iron, tawny browns in farm soils. Noting that developers don't want to find wetlands on their property, soil scientist Art Allen quips, "Brown over yellow, happy fellow. Black over gray, run away!"

Where the happy colorful soils meet the construction-stopping wetland soils, or in wetlands that are not wet all the time, the situation gets tricky; colors, textures, and depths become extremely important. Here, the soils may not have as much organic matter in them, so they are considered mineral soils. When the water sits long enough, oxygen becomes depleted and the iron oxides—which create the rusty-red colors in dry soils—are transformed into lighter orange and yellows, or gray. Wetland scientists have to look at what percentage of the soil near the surface shows orange spots (called mottles) and what percentage is gray to decide if it the area has been wet enough, long enough, to have developed into a hydric (wetland) soil. (Interestingly, according to Tom, in areas where wetlands have recently formed, because of changes in hydrology, the soil can show signs of hydric soils within three years.) Other indicators of hydric soils include bits of black organic matter washed downward into the subsoil, and purple-black nodules of manganese. Even the stinky rotten-egg smell that often rises from a wet area is an indicator of wetland soils—this is hydrogen sulfide, created when oxygen is used up in the soil and the bacteria and other microbes react with sulfur.

The Natural Resources Conservation Service (NRCS, formerly called the Soil Conservation Service) classifies and maps particular soil types as hydric soils for each region of the country. As defined by the NRCS, a hydric soil is a soil that formed under conditions of saturation, flooding or ponding long enough during the growing season to develop anaerobic conditions in the upper part. Using NRCS soil maps (which you can visit from a distance via the most wonderful Wetland Mapper tool, https://www.fws.gov/wetlands/data/mapper.html) to locate areas of hydric soils is an excellent first step for anyone attempting to discover not-so-obvious wetlands — even wetlands that were filled or drained long ago.

Wetlands that were filled long ago are actually quite sought-after in some areas, as their restoration can be the ticket to filling an existing wetland in the way of a building project. Trading the destruction of one marsh that stands inconveniently in the way of a highway, pipeline, or shopping mall, for the reconstruction of a wetland filled or drained before the value of these rich areas was recognized is standard practice in most of the country. Mitigating the impacts in one place by remaking the marsh elsewhere — it is a maddening trade-off, fraught with ecological, ethical, and economic challenges.

Mitigation as Compromise

The relocation of the South Beltline Highway around Madison, Wisconsin, in 1985 meant unavoidable wetland loss ("no practicable alternative," in regulatory parlance). Highways and wetlands have a long and antagonistic history because roadways, being long and linear, are particularly difficult to route in and around wetlands. The Wisconsin Department of Natural Resources removed its opposition to the South Beltline project only when the mitigation — a combination of reducing losses and restoring former wetlands — was deemed sufficient, and alternative alignments were considered too disruptive of surrounding communities.

The Beltline would cut through Upper Mud Lake Marsh on either side of a new bridge crossing the Yahara River. To compensate for the resulting loss of twenty-two acres of sedge meadow, the Wisconsin Department of Transportation (WisDOT) would restore a portion of marsh previously filled for an old foundry sand dump. At this point in time, freshwater wetland mitigation activities had not progressed much beyond wildlife pond creations (i.e., "duck doughnuts"). In previous years they may have dug out the sand, let it fill with water, and called it good when the ducks landed, but in 1985 Wis-DOT found itself on the cutting edge of wetland restoration techniques. This was Madison, after all, with its history of protest marches on controversies of every stripe, so the roadway had been a very contentious project. Restoring the lovely sedge meadow that stood in the highway's path became the end game. Taking its cue from Florida restoration projects, WisDOT proposed an innovative technique using salvaged marsh surface from wetland slated to become roadway. Between 1985 and 1986 the foundry sand in the former marsh was removed and the donor marsh soil, complete with sedge meadow seed bank, was laid down. With a nod to old-fashioned wildlife management, WisDOT did throw in three duck doughnuts for good measure.

Catherine Owen Koning and colleagues (Owen, Carpenter, and DeWitt 1989) evaluated the project two years after its completion in 1986. Comparing the restored site to an adjacent, undisturbed reference area, they found vegetative similarities but greatly altered soils and hydrology at the restoration. The major shortcoming of the restored area was incomplete removal of the sand, on top of which was laid a paltry six-inch layer of salvaged marsh surface. The soil layers of the undisturbed wetland consisted of a top layer of characteristically dark peat for about three feet, underlain by marl (loose sedimentary soil high in calcium), in turn underlain by gray clay. This thick layer of peat produced a steady moisture regime in the original, "reference" sedge meadow, a preferred condition for sedges. In the restored area, the thin layer of peat undergirded by sand produced extremes of dry and wet conditions that proved too harsh for most of the native plants.

In 1991 Sharon Ashworth found herself five years removed from project completion evaluating plant community changes in the attempted sedge meadow restoration (box 7)—just how close did the restoration come to the real thing? As it turned out, pretty close (Ashworth 1997). The replacement feature was indeed a wetland and dominated by plants typically found in the sedge meadows of southern Wisconsin. However, because of the disparity in the thickness of peat layers and hydrology, the plant species' diversity, distribution, and cover in the restored meadow were different than those found in the undisturbed wetland. The patchy nature of the soil profile in the restored wetland—a section of thick peat here, a section of thin peat underlain by sand there—meant patchy vegetation. Desirable sedges and

Box 7. Contemplating Mitigation on a Hot Day

Author Sharon writes: Pave it. Just pave over the whole damned thing—see if I care. But I did care, which was the problem. I am standing in the middle of compromise, that murky intersection of want, need, guilt, and "if the nature lovers insist." It was one of those places where capitulation to modern life is assuaged by making facsimiles and creating novel replacements of the real thing. And right now that intersection is really getting on my nerves. After traipsing across the soggy sedge meadow, I had entered the newly built wetland and fallen in the same hole for the third time while surveying the vegetation. After several humid, hot hours being close to nature, one green plant began to look like any other green plant. But I had signed up to report on the state of compromise: whether this replacement wetland could pinch-hit for the original player. The original player being under several hundred tons of concrete, the replacement would just have to do no matter what I find. No one ever visits anyway—no trails, no picnic tables, no scenic overlooks; just marshy ground between the "our city's economic well-being depends on this" highway and the "vital to our city's quality of life" lake. A pox on both their houses.

I am tired, physically and mentally, of the compromises—and feeling guilty for wanting to pack it in and drive, not bicycle, to the nearest air-conditioned grocery store for a beer and a bag of chips coated in decidedly nonorganic orange dust. Maybe a vacation someplace where they don't care so you have to throw your aluminum can in the trash. Anyway, what does it matter if the new wetland is covered with as much sedge as its esteemed, "pristine" counterpart?

I really don't know; I am here to find out. I rest. The redwing blackbirds trill their ownership rights, a sedge wren plays hide-and-seek. I can see the threatening leading edge of the reed canary grass invasion—the boundary between diversity and monotony. The plants come back into focus. I can distinguish the slender, blue-green stems of bluejoint grass from the arcing, slightly deeper green blades of the sedge. I care. Reparations for damage done do matter.

I rise and wade back into the compromise. I can hear the new road on which I will drive back to town and know I will bike to work tomorrow; I can taste the beer I will have later and know I will recycle the bottle. I live in both houses.

I hope I miss the hole this time.

grasses were found on thicker peat and, unfortunately, invasive willow on thin peat. (Willows don't mind extremes of wet and dry, while sedges and bluejoint grass prefer a more stable home.)

As time progresses, it may be that these neighboring sections of wet meadow take different vegetative paths. Will the restored area be considered a failure if it is overrun by an invasive species or if it does not quite match its undisturbed counterpart? What does a restored wetland need to be in order to be declared a success? These are the sorts of questions that restoration science is tackling with greater and greater sophistication.

Wetland restoration is now a full-fledged industry, mostly driven by mitigation, which employs thousands of consultants and government professionals. Wetland science and restoration science have matured into academic disciplines whose research informs mitigation activities as well as conservation and management of our remaining wetland resources. The Society of Wetland Scientists, formed in 1980, has nearly three thousand members in more than sixty countries and has established a professional certification for wetland scientists. Reports on the status of wetland losses and gains have been regularly produced since the 1990s, and the National Wetlands Inventory has been mapping wetlands across the United States since the mid-1970s. No other ecosystem has what amounts to a national tracking system, national statutory protection, and certified professionals dedicated to its protection and rehabilitation. All this is a testament not only to the ecological importance of wetlands but to the fascination they engender in those that seek to understand and protect these ecosystems.

Status of the Science

The fundamental assumption of no-net-loss is that wetlands can be created which function equivalently to natural wetlands.
—KATIE HOSSLER et al., "No-Net-Loss Not Met for Nutrient Function in Freshwater Marshes"

Trying to put hamburger back on the cow.
—ROBERT ASHWORTH, Groton Connecticut Inland Wetlands Agency, describing efforts of wetland mitigation

With burgeoning restoration and creation activity brought on by mitigation and no-net-loss policies, the National Academy of Sciences (NAS) undertook to review and assess replacement wetland ecosystems. What they found was not promising.

> The goal of no net loss of wetlands is not being met for wetland functions by the mitigation program, despite progress in the last 20 years. (National Research Council 2001)

On paper, an average of approximately 1.78 hectares (4 acres) of wetland restoration or creation was required for every permitted loss of 1 hectare (2.5 acres). The net result should have been a win for wetland acreage on the ground (Turner, Redmond, and Zedler 2001). However, the authors of the NAS report found that just because a permit is given does not mean a functioning wetland is built in exchange for one lost. Only 21% of mitigation sites met ecological equivalency tests with regard to functions lost; the permit system was allowing 80% net loss of wetlands.

For example, in Massachusetts, nearly 22% of permitted projects did not even attempt to mitigate wetland losses, and 38% of attempted projects did not produce anything that qualified as a wetland (Brown and Veneman 2001). Most of the wetlands that were built were smaller than required and did not replicate the plant community of wetlands destroyed. Pennsylvania and Indiana fared somewhat better, with studies in those states reporting a mitigation success rate of 62% (Cole and Shafer 2002) and 64% (Robb 2002), respectively.

The NAS report presented an incredibly important perspective on wetland restoration and mitigation science and policy. It highlighted that mitigation policy must be much more than bean counting—accountability not only for wetland acres but for wetland functions. Indeed, the US Fish and Wildlife Service found in its *Status and Trends of Wetlands* report that over the period of 1998 to 2004 there had been a net gain in wetland area (Dahl 2006). But the report was careful to note that there were no indications of a net gain in wetland functions. The report also found a disturbing trend: a shift in wetland types. Coastal wetlands and forested wetlands declined and freshwater wetlands increased—particularly, open pond areas. The fundamental assumption of no net loss is that restored or created wetlands will

replace wetlands destroyed, that they will be of the same type and function-
ally equivalent.

Subsequent studies continue to document the shortfall of mitigation
techniques and policies. Constructing ephemeral ponds (vernal pools — see
chap. 6) as critical habitat for amphibians appears to be particularly tricky,
with constructed ponds typically not drying out as needed, and as a result,
harboring predators — bullfrogs and fish dining on salamander larvae (Vasconcelos and Calhoun 2006; Denton and Richter 2013). A review of the literature comparing the desired-goal wetland (reference wetland) with the
restored or created wetlands of all types worldwide suggests we are far from
our aspirations.

> In many wetlands . . . ecosystem services may not be fully recovered even
> when wetlands appear to be biologically restored. If markets for ecosystem
> services and mitigation offsets from restored or created wetlands are used
> to justify further wetland degradation, net loss of global wetland services
> will continue and likely accelerate. (Moreno-Mateos et al. 2012)

The loss of wetland services — the functions or processes that occur
naturally in wetlands — seems most evident when a local birding spot or
frog-breeding pond goes silent, when boats are needed to travel local roads,
when the sea takes the first row of beach houses, or when lakes go green.
Less noticeable is the loss of carbon storage.

Many studies comparing restored or created wetlands to natural, or
reference, wetlands still focus on wetland structure, flora, and fauna. More
recently a critical eye is being turned to comparisons of biogeochemical
functioning. A 2011 Ohio study found that created wetlands store 80% less
carbon in the soil and process 60% less nitrogen through denitrification
when compared to natural wetlands (Hossler et al. 2011). These estimates
are in line with a study that took a worldwide perspective on how restored
and created wetlands stack up to reference wetlands on a number of features (Moreno-Mateos et al. 2012). While the hydrology of the wetlands
returned to natural conditions relatively quickly, within the first five to ten
years, biological structure (plants and animals) recovered only 77% of the
original wetland (reference) value — even after one hundred years, in the
case of the handful of wetlands with records over that length of time. The

biogeochemical recovery (carbon storage and nitrogen and phosphorous cycling) of wetland function lagged as well, achieving only 62% of reference value after twenty to thirty years.

> Restoration performance is limited: current restoration practice fails to recover original levels of wetland ecosystem functions, even after many decades. If restoration as currently practiced is used to justify further degradation, global loss of wetland ecosystem function and structure will spread. (Moreno-Mateos et al. 2012)

The Carbon Question—Including Carbon Sequestration in the Mitigation Compromise

You may not think you can observe carbon storage in wetlands, but it's there, in the plants and in the soil. Wetlands are key players in the carbon cycle and thus assert a role in the changing climate, storing up to 35% of all terrestrial carbon and producing up to 75% of all nonanthropogenic methane (Artigas et al. 2015; see box 8). As mentioned in chapter 7, coastal wetland systems are great warehouses of carbon storage and the unsung heroes of carbon sequestration. They and the vast northern peatlands garner most of the climate science community's concern as the former is flooded by sea level rise and the latter dries out as temperatures rise. As for the freshwater wetlands in between, there are few estimates of carbon storage capacity, and what few exist vary widely.

Uncertainty in estimates of carbon storage, however, does not dampen scientific or bureaucratic enthusiasm for counting wetland restoration as part of a global strategy for climate change mitigation. Carbon stored in wetlands can now be bought and sold on the carbon market. Say you own an airline company and want to advertise as an "environmentally sensitive" airline company. Of course, each flight produces literal tons of carbon dioxide, but you can offset the carbon dioxide produced by paying for carbon sequestration. Coastal wetlands store carbon, so you buy carbon sequestration credits from preserved or restored wetlands. A third party verifies the carbon storage capabilities of the wetland and determines the number of carbon credits a particular wetland or wetland complex is worth. Proponents of the carbon market highlight the funds generated by the market for wetland preserva-

Box 8. Basics of Carbon Sequestration

There are two greenhouse gases that we need to concern ourselves with as they relate to wetlands: carbon and methane.

Wetlands store carbon from the atmosphere in plants and in soil. Plants growing in wetlands photosynthesize, removing carbon dioxide (CO_2) from the atmosphere around us and retaining, or sequestering, the carbon molecule in stored sugars and plant tissues. When a plant dies, that carbon is trapped, or sequestered, in the soil because wet conditions inhibit decomposition (water in the pore spaces of soil displaces oxygen, which is necessary for decomposers to break down organic matter). Any carbon imported with sediments from the upland is also trapped in the wetland soil.

A wetland that sequesters more carbon than it releases is called a carbon sink. Carbon stored in buried plant material is released over time as plant tissues slowly decompose; the balance of carbon sequestration versus release depends on the type of vegetation and how long the soil remains saturated. Carbon release is boosted if the organic material and its decomposers are exposed to oxygen as water tables drop—because of drought or aquifer depletion, for example. Conditions that deplete soil moisture—such as installing drains for agriculture, or climate change—can turn a carbon sink into a carbon source.

Wetlands also produce and release methane when microorganisms decompose organic matter. When soils are highly reduced (very little oxygen)—a condition created by continual, long-term saturation—certain bacteria (methanogens) in the soil convert CO_2 to methane, which is then released into the atmosphere.

The question is, just how much carbon dioxide do wetlands remove from the atmosphere on balance with carbon dioxide and methane releases? Whether a wetland is a carbon source or sink has to do with the types of plants and how much plant matter is standing in the wetland, how saturated the soils are and how much sulfate is available—salt water versus fresh water. The sodium, chloride, and sulfate in salt water inhibits microbial decomposition, reducing carbon emissions. Methane emissions also take place less often in salt water because, in the lineup of chemicals available to microorganisms for decomposition, sulfur is worth more than the carbon found in plant matter, so sulfur is used first. When microorganisms do not have other options for obtaining energy, they turn to carbon; and when they do, methane is produced. With so much sulfur available in salt water, there is little need to depend on carbon.

tion and restoration. You can't stick a carbon gauge in the soil like you can a moisture gauge so creating and verifying a carbon-market-ready wetland is complex, time-consuming, and imprecise.

As with their dryland counterparts, wetland ecosystems remove carbon dioxide from the atmosphere, but reverse course if disturbed. Disturbance exposes soil microorganisms to oxygen, jump-starting their ability to decompose dead plants, which in turn releases carbon dioxide and converts carbon sinks to carbon sources. The sequestration and release of greenhouse gases by wetlands depends entirely on the soil, hydrologic, and vegetative characteristics of an individual wetland. The base of each system is the underground microbial communities that break down and consume organic matter and drive greenhouse gas cycling. These communities in turn are affected by the changing climate, and their response dictates whether a wetland serves as a net carbon sink or source (Artigas et al. 2015).

Dr. Amy Burgin, University of Kansas, is one of those scientists who just truly enjoys spending time simply chatting about wetland biochemistry. She is also very adept at scaling up talk of nitrogen, carbon, and sulfur to discussions of mitigation policy and the future of wetland restoration. Amy's research is at the cutting edge of questions regarding freshwater wetland contributions to greenhouse gas cycles. Before embarking on her latest research in freshwater systems, Amy did spend time in coastal, carbon-sucking wetland systems and came away with a new appreciation for the nonchemical elements of salt marshes.

"I was not prepared for the number of things that could kill you," Amy explains about her work in a North Carolina coastal wetland. She had agreed to collaborate with a colleague on a project in a restored salt marsh—sight unseen, and in July. Working in Ohio at the time, Amy had never worked in coastal systems. Excited to get out into the wetlands upon arriving in North Carolina, she took with her a field crew consisting of another young professor, a graduate student, a couple of undergraduates, and a field technician. Rather than carry around supplies in the insane heat, they left the coolers out on the side of the road and hiked into the wetland. She soon realized it was not going to be a walk through one of those sunsets on the salt marsh postcards. Within five minutes of being on site, her undergraduate student stepped over a timber rattler. They pressed on without incident, but upon return to the road they discovered their mauled coolers—evidence of yet another denizen of southern coastal areas, the black bear. The field crew was undeterred, but before embarking on another foray into the wetland the next day, Amy mapped out the path to the nearest hospital, an hour and a half away.

Back in the Midwest, in the relative safety of the agricultural landscape, Amy and her colleague and husband Dr. Terry Loecke set out among wither-ing cornstalks to monitor gas exchanges in a developing wetland. The corn-field is now the Great Miami Wetland Mitigation Bank, part of Five Rivers MetroParks, northwest of Dayton, Ohio. The 114-acre farm was slated to become a landfill, but Michael Enright, conservation biologist for the Five Rivers MetroParks, recognized a restoration opportunity. Michael is a former student of professor emeritus Dr. Jim Amon of Wright State Uni-versity, the fen expert we met in chapter 4. Having been familiar with Jim Amon's classes and restoration activities at the Beaver Creek Wetlands on the other side of Dayton (see chap. 9), Michael sought a way to convert the farmland into something other than a landfill. He convinced Five Rivers to buy the property and create a wetland mitigation bank. Rather than conduct mitigation activities themselves, developers can buy credits from banks such as this one if their projects impact wetlands negatively. Money from the sale of mitigation credits is then used to preserve, enhance, and restore addi-tional acreage. The Great Miami Wetland Mitigation Bank is one of only a few publicly owned wetland mitigation banks in the country.

Five Rivers highlights the mitigation bank as a place to hike and watch birds, specifically advertising avian visitors not typically found in an Ohio backyard. Also not typically found in an Ohio backyard are soil sensors and gas-sampling chambers. For Amy, her colleagues, and her students, the corn-field mitigation bank is a place to monitor the good and bad of greenhouse gas exchange. They hope to answer a basic question: How, when creating or restoring a wetland on the landscape, can you minimize wetland gas emis-sions and maximize carbon storage?

In the farm fields of the Midwest, one needs to be particularly cognizant of a third greenhouse gas: nitrous oxide. Agricultural soils are high in nitrate from added fertilizers. Denitrification—a microbial process in soil that helps decompose organic matter—converts nitrate, a pollutant, to nitrogen gas (N_2), which escapes into the atmosphere. Nitrogen gas is not a green-house gas and is in fact the dominant gas in our atmosphere. However, incomplete conversion of nitrates produces nitrous oxide, a greenhouse gas more potent than either carbon dioxide or methane. As with carbon dioxide and methane production, the degree of nitrate reduction to N_2 or N_2O is dependent on the amount of oxygen in the soil, which in turn is dependent on soil saturation (Burgin et al. 2012). Generally speaking, the more saturated the soil the less oxygen available, which results in greater releases of nitrous oxide. However, things are never that simple—the *duration* of saturation and the temperature of the soil also influence oxygen fluctuations (Jarecke, Loecke, and Burgin 2016), which in turn influences production of nitrous oxide. To achieve all the steps in the process of denitrification, a wetland needs to have both anaerobic (no-oxygen) and aerobic (oxygen-present) phases (see chap. 2).

Production of the potent greenhouse gas nitrous oxide is of course considered a wetland *dis*service. On the other hand, nitrogen removal is a long touted service of wetlands. Amy, who is originally from Des Moines, Iowa, points out that her hometown spends approximately $7,000 a day to remove nitrogen at its water treatment facilities—a burden that wetlands could soften. This is just one of the many reasons we should care about the biogeochemistry of wetlands beyond establishing water, plants, and wildlife. But Amy still has a hard time explaining the need to care about wetland biogeochemistry. While at the federal level there is more openness to wetland-climate and water-quality issues, biological indicators such as wetland plants and animals are still the focus of the boots-on-the-ground regulators and managers. Often, she and her team don't even talk about carbon; they focus on plant growth, hydrology, and birds. But biogeochemistry is an equally important indicator of wetland restoration success. If the biogeochemistry of a wetland is reestablished, you are well on your way to a sustainably functioning system.

Alterations and Adjustments: The Future of Wetland Restoration Science

Alterations

We are now faced with a new regime: how to restore wetlands while the environment changes under our feet and over our heads. The consequences of global climate change for wetlands are varied and uncertain. There will be fundamental changes in wetland hydrology—water quality as well as quantity—as some areas of the globe experience longer dry spells and others experience more intense storms and flooding. Higher average temperatures will alter plant and animal life cycles (phenology) and accelerate the spread of exotics. Some plant and animal species' home ranges will shift with warming temperatures, and some will not make the adjustment.

Climate change in the United States is expected to make dry areas even drier and stormy areas much wetter. The Midwest and far West of the United States will see thirsty soils going longer periods without the relief of a good soaking rain. In the Northeast, by contrast, wetter conditions will prevail, although less frequent snowfall and big downpours will be interspersed within longer periods of drought; higher temperatures will mean increased evaporation and evapotranspiration (Brooks 2009; US Global Change Research Program 2014). Consequent changes in water depth and hydroperiod (wet and dry periods brought on by seasonal changes) could transform wetland plant and animal life. Some of the most vulnerable freshwater wetlands are likely to be rainwater-fed, ephemeral wetlands. Depending on individual circumstances and locations, we could see these ecosystems drying up earlier in the year or disappearing altogether. In those pools that remain, the list of macroinvertebrates will be restricted to the shorter-lived creatures. Amphibians and reptiles may be evicted earlier in the year and need to travel much farther to find suitable quarters for the remainder of the summer.

Additionally, higher temperatures translate to warmer water temperatures in small woodland ponds, causing a shift in amphibian life cycles. Frogs around Ithaca, New York, are starting to call nearly two weeks earlier than they did at the beginning of the previous century (Gibbs and Breisch 2001). In Wisconsin, salamander numbers are expected to peak earlier in response to warmer temperatures, and drier summers will shorten flood periods and alter the timing of juvenile emigration from the ponds (Donner et al. 2015).

Floodplain wetlands may experience higher flood levels and excessive erosion from the surrounding landscape as a result of increasingly severe storm events; yet these same marshes and swamps may be parched in summer months as time between events lengthens. Data from streamflow stations around the Great Lakes show an earlier occurrence of spring peak flows, as a result of earlier melting of snowpack (Janowiak et al. 2014).

Riparian forests are particularly adapted to, and therefore sensitive to, annual and seasonal changes in water table. Flooding regimes determine the annual composition of pioneer tree seedlings (Dixon 2003) — those species that sprout on freshly scoured sandbars. Changes in the timing and depth of floods, along with temperature changes, may affect which tree species will survive in our floodplain forests. As for our bog ecosystems, all their adaptations to harsh, cold conditions may be their demise, as cold-adapted species such as black spruce and tamarack trees are expected to decline in number as temperatures increase (Janowiak et al. 2014).

Even the groundwater that seeps into fens and other wetlands is not immune from the changes wrought by warmer temperatures. Demands on aquifers increasingly divert groundwater to human use, leaving inadequate amounts for peat formation in fens and bogs. Groundwater flow to calcareous fens and springs along the Minnesota River are reduced when wells are pumping at peak rates. Urban water demands in Madison, Wisconsin have caused some springs and fens connected to the shallow aquifer to dry (Quentin Carpenter, pers. comm., April 20, 2016). The fens of the University of Wisconsin Arboretum in Madison, originally surveyed by John T. Curtis in his classic book *The Vegetation of Wisconsin* (1959), are highly degraded, their groundwater supply altered and the native vegetation replaced by buckthorn and reed canary grass.

Those who practice restoration not only need to tweak plans in an attempt to accommodate changing weather patterns, but they may also need to adjust their goals and vision of what restoration means and what future functions and values wetlands may provide. For Suzanne Pittenger-Slear of Environmental Concern, that simply means a wetland restoration completed as designed, plants where they should be, and a functioning habitat.

Adjustments

So, what did happen to the South Beltline wetland on the outskirts of Madison? UW–Madison wetland researcher Quentin Carpenter, who along with Catherine Owen Koning studied the wetland in 1988, says that the restored areas near the roadway are now mostly cattail. The years he and the authors of this book spent in those wetlands were relatively dry; but, since that time, more years than not have been very wet. This is consistent with predictions of climate change for Wisconsin—wetter, warmer, with more frequent storms. So, rather than the sandier bits of the restoration suffering low water tables, allowing invasives like willow to proliferate, the area has been extremely wet, allowing cattail to invade and thrive over most of the wetland. Nearly twenty-eight years later, it remains to be seen if this cutting-edge restoration can achieve the goal of replicating the "original" wetland.

Although wetland professionals and amateur enthusiasts never truly replicate lost wetlands or complete the ecological equivalent of time travel by restoring land to "predisturbance" conditions, it remains a prevalent goal, if only on the paper permit. In the context of a changing climate, such bureaucratic attempts and visions begin to seem especially quaint.

Before its recent reincarnation, the Great Miami Wetland Mitigation Bank in Ohio had ceased to be a wetland more than a hundred years ago. "Who knows what you're actually trying to restore it to," says Amy Burgin about the land formerly stocked with corn and soybeans and doused with agricultural chemicals. Beneath the surface are miles of tile drains, installed to move water off the land so it could be farmed. Surrounded by additional agriculture and development, there is little opportunity for plants and animals to disperse and repopulate the new wetland. Yet wetland mitigation banks such as this one are a major engine of the restoration industry. Mitigation is a business and the physical nature of wetlands is relatively easy to replace—or is at least deemed easily replaceable. "The no-net-loss policy is really at the heart of this focus on physical structure of a wetland," Amy notes. "[The sentiment is that] I can make one here that will look just like it, so why can't we just do that everywhere and rearrange the landscape however it is convenient for us?" She responds to this rhetorical question: "Yeah, you can make the new wetland look like the one that was destroyed—the vegetation, the animals, those things come back quickly. But what you lose is the biogeochemistry, and that's increasingly important as we have these more global and regional problems to focus on, like water quality in the Mississippi River basin or global climate issues."

Armed with nearly five decades of practice and ever-advancing knowledge about wetlands, we may now be able to restore and manage wetlands with an eye to preparing these ecosystems for an uncertain future (i.e., designer wetlands). Amy and Terry agree that this is where wetland science is headed. Scientists and engineers already have a great deal of experience with designing wetlands for water treatment, and Amy sees that particular wetland restoration goal becoming more prevalent; but a new direction is evident in discussions of frequent flood events, sea level rise, and carbon markets. Terry points to coastal wetland restorations that are now designed in response to sea level rise (see chap. 7) and suggests that floodplain wetlands will need to handle greater extremes of swelling floodwaters. As for freshwater, inland wetlands, the Ohio mitigation bank Amy and Terry are working in is a potential model for designing climate-friendly freshwater wetland restorations.

The "Tyranny of Small Decisions"

Wetland science and wetland appreciation continue to evolve, but wetlands still get in the way of what we as a society deem more important: a faster commute to work, commercial development to increase city tax bases, new houses and condos—the tyranny of a thousand small decisions (Vileisis 1997). And so a permit is given for 4.6 acres of wetlands filled for a new highway, a mere half acre filled for a hotel parking lot two towns away, eight acres for a natural gas pipeline in the next county.

> The net wetland loss was estimated to be 62,300 acres between 2004 and 2009, bringing the nation's total wetlands acreage to just over 110 million acres in the continental United States, excluding Alaska and Hawaii. (Dahl 2011)

So states the latest assessment of wetland loss and gain as reported in the most recent release of the series *Status and Trends of Wetlands in the Conterminous United States* from the US Fish and Wildlife Service. The report goes on to say that, while the rate of wetland gain through "reestablishment" has increased, so has the rate of wetland loss—and by a much greater factor: 17% gain to 140% loss over the previous measurement period, 1998 to 2004 (see box 9).

But what if all those small decisions go the other way?

Brett Amy Thelen of the Harris Center for Environmental Education in New Hampshire (see chap. 6) gives an example from 2008: a road-widening proj-

Box 9. What Remains

Ninety-five percent of all remaining wetlands are freshwater, and approximately one-half of that acreage is forested wetland, with the remaining acreage in shrubs (26%), emergent wetlands (18%), and ponds (6%). Of the 5% of wetlands that are saltwater wetlands, the majority (67%) are the familiar coastal salt marshes, with nonvegetated (21%) or salty shrub wetlands (12%) making up the rest of that 5%. Our biggest wetland losses, as a percentage of the total, have been our coastal salt marshes, followed by freshwater forested wetlands. We lost 84,100 acres (34,000 ha) of salt marsh from 2004 to 2009—three times as many acres as the previously measured period, 1998 to 2004. We lost 633,100 acres (256,200 ha) of freshwater forested wetlands from 2004 to 2009—a somewhat slower rate of loss from the previous study (Dahl 2011).

ect in Keene that went right through a vernal pool. The contractors did not know the pool was there. Brett and some volunteers investigated the site while the project was under way and were able to amend the road contract to remove riprap from the pool and clean up the construction debris, reducing the permanent damage. In another instance, a vernal pool inventory volunteer found a pool in an area where a power company was replacing utility poles. The heavy equipment would access the site through the vernal pool, but the utility company was happy to alter the equipment route once the volunteer explained the nature of his concern.

Even in the South Bronx of New York City, where wetlands have long been replaced by concrete and garbage, there are possibilities for redemption. Dave Kaplan was working for the New York City Department of Parks and Recreation in 2002 on what some might deem a hopeless project. The dump Dave found himself in was real: an abandoned concrete factory on the banks of the southern end of the Bronx River—a site earmarked for a small tidal salt marsh restoration project. The salt marsh and adjacent upland were to be part of a much larger endeavor bringing a series of green spaces and parks to blighted areas along the river. The Bronx River has a long-held reputation for being more sewer than river, and Dave describes the site as filled with "hypodermic needles, homeless people, and junkyard dogs." The city's combined sewer overflows during storms (overflowing sewer systems are combined with regular stormwater) meant the park department's crew and contractors encountered any and all manner of things people flush down toilets.

A huge portion of the restoration effort — and the budget — was spent on soil and concrete removal. The planned salt marsh location was the former rinsing area for concrete trucks, so the now hardened rinse water had to be trucked out, along with thirty-two thousand tons of contaminated soil (Kimmelman 2012). Dave would meet the contractor hired to do the removal each day, and at the end of the day they would lock the gates surrounding the site. Now, this is metropolitan New York City, and, of course, working in hidden, out-of-the-way places, there were always jokes about "finding a dead body." One morning, when the restoration crew opened up for work at 6:30 a.m., they found a trash bag at the bottom of the riverbank that had not been there when they closed up the night before. It *was* a dead body that had been dumped overnight. (Dave recalls a handful of people he knows who work in the parks and on the margins of the city who have indeed encountered bodies. It should be noted here that Dave currently works along the Gulf coast.)

The situation along the Bronx River is about as bleak as one can imagine, but now this dump, renamed Concrete Plant Park, is part of a network of rehabilitated green spaces along the Bronx River and is frequented by neighborhood residents.

> Park by park a patchwork of green spaces has been taking shape, the consequence of decades of grinding, grass roots, community-driven efforts. For the environmentalists, educators, politicians, architects and landscape designers involved, the idea has not just been to revitalize a befouled waterway and create new public spaces. It has been to invest Bronx residents, for generations alienated from the water, in the beauty and upkeep of their local river. (Kimmelman 2012)

One could easily imagine a blighted, wet, weedy patch being given up for lost, or stripped to lay down a fresh patch of concrete. But, as the next chapter shows, sometimes people make small decisions that challenge the status quo.

CHAPTER 9

Beauty, Ethics, and Inspiration

A land ethic of course cannot prevent the alteration, management, and use of these "resources," but it does affirm their right to continued existence, and, at least in spots, their continued existence in a natural state. . . . It is inconceivable to me that an ethical relation to land can exist without love, respect, and admiration for land, and a high regard for its value. By value, of course, I mean something far broader than mere economic value; I mean value in the philosophical sense.
—ALDO LEOPOLD, *The Land Ethic*

Noting that his favorite wetlands, the Atlantic white cedar swamps, have declined precipitously and don't seem to be growing back in restored areas, Rob Atkinson says, "One of the things we see in a changing planet is loss, of ecosystems and species. Wetland scientists always have to cope with this sense of loss." Telling our stories of tribulations and triumphs, discoveries and delusions, may be one way to grieve these losses. By celebrating the life of a shrub swamp or a marsh, by committing to educate others about the joys of wetland exploration and about the intricacies of their workings, perhaps we can also stem the tide of loss.

Many of the narrators in this book have spent their entire working lives telling scientific stories using measurements and data. In retirement they take care of the wetlands that not only nurtured their careers but fed their souls. In Rhode Island, Frank Golet continues to visit Diamond Bog, offering expertise and manpower to management efforts. In New Jersey, Mary Allessio Leck works with the Abbott Marshlands, developing educational materials and bringing high school students out to the tidal marsh. In Maine, Ron Davis leads naturalist walks along the boardwalk in Orono Bog, and in Ohio, Jim Amon still works to protect the uplands around the Beaver Creek Wetlands fen. This impressive commitment springs out of nothing less than love. As Aldo Leopold notes in the chapter epigraph, we love what we know and admire, what gives us joy and stimulates our intellect.

Although many scientists, explorers, and educators devote a lot of their lives to wetlands, the general public may still lag in their appreciation. Many people would profess to have an aesthetic preference for nature over the built environment, but they tend to like nature a bit tamed—neat and tidy, with some open water and a boardwalk, please. A landscape perceived as aesthetically pleasing is more likely to be appreciated and protected regardless of its ecological value (Gobster et al. 2007). However, function and appearance do not necessarily correlate.

When most people view a natural area, they may quickly grasp its importance for wildlife; but few are likely to notice ecological functions such as water filtration, carbon sequestration, flood control, or shoreline stabilization. These functions arise out of the complex set of connections among species, manifested in a bumpy tangle of branches and stems, itself giving rise to a mosaic of chemical reactions in the soil, water and air. Wetland biodiversity is messy, and people prefer settings more like our ancestral savannahs, natural areas that emphasize short grass with scattered trees and long views. For many established nature reserves, scenic beauty does correlate with ecological value—soaring mountains showcasing rare alpine flowers, or old-growth forests harboring secretive birds—with two notable exceptions: prairies and wetlands (Gobster et al. 2007).

The ugly duckling of landscapes, wetlands are even less attractive if they are not wet. Humans like to look at open water, and they have a hard time accepting wetlands as wetlands if there is no water readily visible. Without standing water or a flowing river, there can be no grand displays of waterfowl calling to compatriots as they circle down to the surface of the pond, no collections of shorebirds to tease apart with a bird guide, no reflections of scarlet-red maple leaves in fall. Without stretches of open water, there is of course a plethora of interesting creatures; but the songbirds, rails, and reptiles that inhabit dense vegetation often remain hidden to all but those willing to venture forth into the muck.

Perhaps the average person needs the data—the science—as well as the humor, the passion, and the excitement that comes from the personal stories of disastrous entrapment in mud, hours lost wandering in bogs, grand encounters with moose, unexpected discoveries of rare orchids or bog turtles.

The folks on the front lines between the unaware layperson and the wetlands themselves are the consultants, the wetland scientists who help companies and private individuals comply with wetland protection laws. Jason Smith, an environmental consultant from Bethlehem, Pennsylvania, knows that landowners don't always understand or appreciate the wetland laws. "For a landowner, having wetlands on their property is rarely viewed as a good thing. The wetland biologist learns to shrug off their misperceptions, knowing the real values of and truths about these wonderful natural resources. Still, after years of thankless work, even the saltiest wetland scientist can feel pretty defeated from time to time. We have all seen beautiful natural areas and wetlands destroyed in the path of progress."

Jason describes a situation in which he was contacted by a young couple who had purchased a large parcel within a beautiful natural area and had begun clearing for a long access driveway that crossed several wetlands and streams. They had not gotten any permits from their town for the work they were doing, and after receiving a report from a disgruntled neighbor, the town stepped in with a stop-work order. Reluctantly, the couple hired Jason to deal with this "nuisance requirement" and get them back on track.

Prior to conducting any fieldwork, Jason discovered that bog turtles might live on this property, and two threatened plant species were listed for the general area. This also did not make the landowners happy, as it represented additional delay and expense for them. The whole project was depressing—sad for Jason to see this lovely complex of shallow pools and stream-laced woods degraded, upsetting for the landowners paying to be told what they could or could not do with their property. Before long, Jason had finished delineating the wetlands and water bodies and obtaining clearances for the listed species. "I found the best possible location for the new driveway. The progress seemed to satisfy the owners, and they were excited to finally move forward with construction. The husband resumed clearing trees for their new driveway and home site."

About a week later, Jason says, he returned to the property. "And as I walked up the newly cleared path, a beautiful purple-and-white orchid which I had never seen before was in full bloom . . . perfectly centered in the clearing for the driveway." It was a rare orchid—but not one listed for protection. Still, it was quite uncommon and deserved consideration. Another unpleasant situation was unfortunately at hand, and a call to the owners was made. "When the wife answered the phone, I explained what I had found. At this point, her patience had come to an end, and she just wanted to march forward with no regard for this special plant. I hung up feeling very disheartened, knowing that one more rare species was going to take another hit. What happened later that day, however, was one of the most surprising and encouraging moments in my career as a wetland biologist." After speaking to her husband, the woman called Jason back. "Her husband was insistent that the driveway be moved and that the orchid be saved! He did not care about the added work or cost to make it happen, either. He wanted it protected, period." She went on to explain that her husband had recently lost his father, who collected and grew orchids as a hobby throughout most of his adult life. Her husband simply wished to honor his father in protecting this wonderful plant as part of this project.

Jason continues, "I read a lot into this decision and have reflected on it many times since that day. It's things like this that keep me interested, knowing that what I am doing actually makes a difference and is worth fighting the good fight! Many people out there either are stewards or wish to be stewards of our environment, and in many cases they only need someone to lead the way down the right path." Thus landowners are transformed from reluctant stewards to emboldened protectors of these sensitive natural communities. In a similar manner, starting from humble beginnings such as the landowners in this story, others proceed to higher measures of caretaking.

Friends of the Bog

One of the first steps in caretaking is the gradual development of a new attitude—one that wetland ecologist Joy Zedler of the University of Wisconsin calls a "wetland ethic," which deepens Leopold's concept of the land ethic. Wetlands, arising within the interstices of land and water, "provide multiple functions that enhance human well-being at rates far greater than their global area indicates." Thus, she argues, people need to enter into a relationship of reciprocity, and "accept obligations along with benefits of wetlands." These obligations include "supporting conservation and restoration of wetland biota and natural functions for posterity. . . . A wetland ethic would foster understanding that protection means more than setting regulations and promising enforcement. A wetland ethic would add voluntary responsibility for ecosystems because they provide services well beyond the small area of earth that they occupy" (J. Zedler 2014).

Out in Ingleside, Illinois, this wetland ethic is very much in evidence. Volo Bog is part of a large state-protected poor fen-marsh complex tucked in among the northwest suburbs of Chicago. Since 1984, Friends of Volo Bog has worked with the Illinois Department of Natural Resources to provide extensive opportunities for members of the public to explore and appreciate this wetland gem all year round. In summer they host the Youth Art Guild, where local artists volunteer to work with young people, helping them explore nature through art. Every fall for almost thirty years, they have lined the boardwalk with pumpkins, hung fake skeletons in the winterberry bushes, and put on a spooktacular "ghost stories in the swamp" event: volunteers hide along the trail dressed up as giant spiders, trash monsters, Auntie Earth, the "bog-man" (or "boogey man," as is he is more commonly known) and even the Lorax; meanwhile, storytellers spin scary tales. Winterfest comes around each February, with snow sculptures, a photo contest, snowshoe treks, live music, and nature crafts. And every week, dozens of docents lead the public on tours of the bog.

Volo Bog inspires this kind of devotion in many volunteers. For example, the mother-daughter team of Julia and Nina Denne of Arlington Heights, Illinois, has been coming to this wetland for more than six years, since Nina was nine years old. "I started volunteering to learn more about the bog, to befriend the plants and learn about the geology of this unique area," Nina writes. "I continue to volunteer at the bog to fuel my thirst for knowledge, to see many of the people I have met who share my interest in the bog, and to inspire others to learn more about wetlands." Nina, a math and science wiz, enjoys the hands-on nature of this outdoor classroom. "It's a way to explore and learn about whatever I like. It is also very relaxing and refreshing to step outside and find peace in nature."

Julia sees how people and nature come together at Volo. "As a mother, I find it extremely important that Nina managed to build connections between learning and the local community. Friends of Volo Bog became our family, and Nina and I feel very lucky to have become part of this inspiring and welcoming community. Nina wants to become a biologist, and she is specifically interested in botany and microbiology. Nina has been helping Stacy and other naturalists teach all-day wetland botany programs, which happen several times a year. Last year, Nina started to mentor a nine-year-old boy who also wants to become a Volo Bog naturalist. She doesn't want to go to college without leaving a knowledgeable and enthusiastic young volunteer in her place." Julia continues: "Volo Bog is also the place that connects Nina and me emotionally. She attends an academically rigorous boarding school, and she is constantly busy. Still, we find time to come to Volo Bog several times a month, and it is our special time together. Every time I say 'Volo Bog,' Nina starts smiling and opens up."

This tight-knit community even tried to hold a camping trip in the bog one year. At the end of a full schedule of Earth Day events, a group of twenty or so volunteers thought it would be fun to sleep out on the boardwalk in the bog. It seemed like a good idea at the time. Botanist and author Linda Curtis (see chap. 2) witnessed the scene: "It was a dark and starry night, cool but not cold. It seemed perfect." The group had a potluck supper, including a large bowl of steaming baked beans and jugs of iced tea.

Strapping sleeping bags and pillows onto their backs and grabbing their flashlights, the volunteers made their way through the black night along the wobbly boardwalk, over the spongy peat, to the wider platform in the middle of the bog. A knotted rope was used to maintain contact in the darkness, as there were no handrails through most of the area. Carefully, the group started out, managing to all simultaneously step on their right foot first, tipping the boardwalk and almost pitching several people into the peat. The group then attempted to coordinate steps to maintain a balanced boardwalk. Bats flew overhead; communication became a game of telephone. "And so it went," Linda writes. "What was normally a half-hour walk became an hour of bumping into each other and being yanked ahead in the dark. Finally, the shrub zone was reached where the plank walk didn't shift side-to-side and the handrails were firmly gripped" (Curtis 2014).

Sleeping bags were unfurled and zipped; flashlights rolled off the platform into the peat, never to be seen again. As soon as everyone had settled in and a few people began to drift off, the iced tea and the beans kicked in. Grumbling, unzipping sleeping bags, stumbling over bodies, folks wobbled their way to the restrooms and back again. It was well after midnight by the time anyone actually slept, fitfully, among the whine of mosquitoes, the snores, and the other gaseous emissions. "At dawn," Linda writes, "aching bones and muscles pulled together, the troop carried their unrolled sleeping bags over their shoulders and returned the long trudge back. There never was another sleep-out" (Curtis 2014).

"Tree-Hugging Pests": The Story of Pheasant Branch

> Never doubt that a small group of thoughtful, committed citizens
> can change the world; indeed, it is the only thing that ever has.
> —Attributed to MARGARET MEAD

Just down the road from Joy Zedler's Madison, Wisconsin, abode, another group of concerned citizens exemplifies the wetland ethic she promotes, dealing with repeated challenges to the integrity of a beloved local marsh.

Pheasant Branch Creek, a tributary of Lake Mendota in the city of Middleton, makes its way through farmland and past suburbs disguised as a ditch. Before it flows into the lake, the stream meanders through a lovely expanse of marshes and meadows. The jewel of the site is a set of springs that sends up flows so strong that even on the coldest Wisconsin winter day they are completely unfrozen. Dozens of rounded boils are formed by the up-bubbling of sweet groundwater, giving it the look of a hot spring. Complex patterns of underground flows make for drastic differences in water chemistry across short distances.

The ecological value of such a complex site is obvious to some of us, but the location of the Pheasant Branch Marsh so close to a large lake, and a recreation-hungry populace made it the perfect spot to dredge for a marina and use the fill for lakeside houses. Or so some thought back in the early

1970s (Tom Bernthal, pers. comm., June 24, 2016). The proposed development roused a small group of neighboring residents to action, forming the Middleton Conservation Committee out of concern for the neglected and abused natural area and stream. Initially the group focused on beautification and clean up, but with the avid leadership of Middleton mayor Wally Bauman, the town acquired two hundred acres of Pheasant Branch marshland from 1972 to 1979. This acquisition prevented the marina and established the Pheasant Branch Conservancy; soon, however, the group discovered that even outright acquisition rarely translates into complete protection.

In 1995, Ann Peckham looked out a window from her house in the Woodcreek neighborhood toward the conservancy one day to see "a bunch of these little flags out there." A bunch of cheerily colored pink plastic flags on wires dotting a natural landscape is rarely a good sign. She went down to city hall to inquire about the flags, and city staff told her of two sewers that would soon run through the marsh. Her response? "I don't think so." And so began the next chapter in the life of the Pheasant Branch Conservancy, with the formation of the Friends of Pheasant Branch (Klubertanz 2016).

As Tom Bernthal, Wisconsin Department of Natural Resources wetland ecologist puts it, "It woke people up to the fact that the conservancy was there in the first place" and a new group of neighbors became active in the conservancy's protection. County government then purchased 120 acres of land adjacent to the north of the conservancy—an area surrounding the springs. A small group of conservancy neighbors began alerting city residents not only to the existence of the marsh, but to the threat posed by sewer lines. Quickly, the group that became Friends of Pheasant Branch—none of whom were ecologists or lawyers—familiarized themselves with the mundane and byzantine language of permitting, the potential ecological impacts of sewer installation, and the law (Klubertanz 2016).

The group's protestations over the sewer line culminated in an unsuccessful lawsuit against the city. The judge ruled that if the Alaskan pipeline didn't hurt the Alaskan environment, a small sewer through Pheasant Branch wasn't going to do any harm (Klubertanz 2016). However, further negotiations with the city and countless meetings resulted in a less destructive route through the marsh. Jim O'Brien describes it as an exciting time, despite their group being thought of as "tree-hugging pests." That was sewer number one. Two years later, when sewer number two was on deck, it was a different story. In the time between the first and second sewer proposals, the Friends of Pheasant Branch had managed to elect a few members to the city council and make allies of other members; the city council voted to relocate the second sewer through a residential street rather than the marsh, despite higher costs.

But devotion to a spring-laced wetland, as with any delicate relation, requires love and constant vigilance because pernicious threats creep in from above, below, right, and left. Invasive species continue to invade year after year and must be controlled. The free-flowing sweet groundwater that nourishes Pheasant Branch Marsh must first pass through adjacent farmland and suburban sprawl, immixing potential contaminants as it flows. Years of observation and enjoyment of the wetland ecosystem led the group to the realization that acquisition *and* legal protection from insidious encroachment were still not sufficient; they needed to add careful management to their tool kit. The Friends of Pheasant Branch moved on to restoring and managing the entire conservancy as a diverse and rich landscape of upland prairie and savannah, lowland meadows and marshes, and groundwater- and surface water–fed streams. Since 1995, the Friends have grown and proved themselves again and again: acquiring grants to plant prairie vegetation, to burn back encroaching shrubs, and to build sediment ponds to intercept pollutants coming from offsite. Today walkers, joggers, birdwatchers, and cyclists use the area. Though a small group of volunteers who can hack down only so many buckthorn bushes and plant so many prairie seeds, they now pack a diverse set of skills, including legal advocacy, raising money, and writing grants — and that may be sufficient.

Behind the Boardwalk: Greetings from Asbury Park, New Jersey

Author Catherine writes: Gathered around a small table, twenty-three people peered excitedly into two gray plastic bins. In one bin, four thumb-size turtle hatchlings scrambled about, and in the other, thirteen palm-size turtles scuttled over, under, and between each other, desperately seeking to escape. Named for the roughly diamond-shaped plates (or scutes) on their backs, this small group of diamond-backed terrapins seemed to know that their freedom was near at hand. Seventeen family groups awaited the chance to release "their" adopted baby turtle into the wild. The greening of steward-ship sometimes starts with one small creature.

We had driven to the New Jersey shore from New Hampshire, two mothers and two daughters in a small car on a warm July weekend. After dis-entangling ourselves from the concrete highways, steel bridges, and indus-trial landscapes of Manhattan and northern New Jersey, we were rewarded with views of miles and miles of salt marsh so green and healthy that it seemed to glow. Famous primarily for Atlantic City, boardwalks, and Bruce Springsteen, the barrier coastal islands and marshes of southern New Jersey harbor some of the best wildlife habitat in the world. After the glaciers re-treated ten thousand years ago, salt marshes gradually formed in the exten-sive bays between the coastal dunes, spits, and barrier beaches at the ocean's edge. On this sultry day, we could see that the area had recently recovered from the ravages of Hurricane Sandy three years earlier; beaches had been replenished, dunes replanted, and bridges rebuilt in hopes of better with-standing the next superstorm.

Our thoughts were not on the natural and human-built infrastructure all around us, however. We were in search of only one creature—one that few people get to see. Every visitor passing over the bridges and causeways to the boardwalks of Ocean City and beaches of Cape May is treated to views of the many egrets, osprey, laughing gulls, and shorebirds who live and hunt in the marshes. But only those who paddle slowly and carefully get to see more secretive species like the diamondback terrapin.

Diamondbacks can hold their own in a beauty contest with painted, spotted, Blanding's, and other colorful turtles. Part of their visual appeal lies in their surprising variability. Terrapin shells can be rusty orange to light gray to brown; their heads, necks, and legs are sprinkled with constellations of small black dots or elaborate patterns of larger dots and dashes set against backgrounds ranging from almost white to dark gray.

All the turtles scrambling in the bins in front of us were females. Thanks to the efforts of the nonprofit conservation organization the Wetlands Insti-tute, of Stone Harbor, New Jersey, these babies had been saved from certain death after their mothers had been hit by cars. Eggs removed from orphan nests are incubated at 86 degrees Fahrenheit (30 degrees Celsius)—at this warm temperature, only females will be produced, because it is the tempera-ture experienced by the developing egg that determines whether the embryo will become a male or a female. "As anyone who spends time around here knows, terrapins are always trying to cross the road, and a lot of them don't make it," Brian Williamson, research scientist at the institute, explained to the animated group of turtle adopters. "Over five hundred turtles are killed by cars each year in this area. We have five hundred and twelve eggs in the incubators right now. We want to replace the females that are killed so there can be a sustainable population in this area."

Young turtles and adult male turtles spend their whole life in the salt water of the bays and marshes along the Atlantic coast from South Carolina to Cape Cod; only the pregnant females ever leave the water, because they must lay their eggs in a dry, protected spot, putting themselves in grave danger on the way to and from their nests. Females lay up to a dozen eggs, which take seventy days to hatch, and most turtles will have two nests each season. Many of the wandering females are hit by cars each year, never having reached their egg-laying destination. Scientists gather these dead or dying turtles from the side of the road and remove the eggs. The eggs that are rescued from the road-killed mothers are incubated, hatched, and kept for one year in the aquariums at the institute before being released into the wetlands of the bay. Surprisingly, the tiny, one-inch-long turtles and the others — four times their size — were all one-year-olds! "Some of them just eat more and grow faster," Brian explained.

Brian then showed a tiny "pit tag," about the size of a staple. "These are injected into the larger hatchlings, so if we catch one again, we can scan it with the handheld laser scanner, and it gives us the number. That's how we know how old it is, where and when it was released." Now the time came for each group to choose its newest family member. Each family had donated $50 to the institute for the privilege of adopting and releasing one of these entrancing little turtles. The tiny turtles went fastest, but no one seemed disappointed with their new relative. Into a blue bucket went each turtle, its transfer to the beach entrusted to wide-eyed eight-year-olds, beaming teenagers, and rapturous adults. After receiving thorough instruction about how to hold a turtle and what to do at the water's edge, the group traveled carefully down the walkway. Along the path constructed of sand and gravel, we passed at least a dozen black-wire cages, protecting terrapin nests discovered by institute interns on their daily rounds. "We have found over eighty nests this year," Brian said. "We continue to monitor them, to see if the hatchlings emerge on their own or if the nest gets dug up and eaten by predators, despite the cage that protects them."

Finally, the big moment. One at a time, my daughter, Mia, and her friend Elyse removed their turtles (named "Terry Pin" and "Coral") from the buckets, leaned over, and gently let them go, watching as the little turtle ladies paddled a few feet into the water. Immediately, both turtles turned around and tried to walk up onto dry land. "They aren't ready to go!" one woman commented. Mia picked up Terry and let her go closer to the salt marsh grasses under the dock, where the little turtle disappeared into the watery darkness. Elyse gently turned Coral around to face the water. Hesitating only a moment this time, she swam off.

We spent the next half hour picking our way carefully along the water's edge, watching the tiny turtles navigate the wettest parts of the salt marsh, and running back and forth on the dock, trying to find the larger turtles in the open water. As the larger ones popped their heads up all over to take a look at the big wide world, they seemed to be enjoying the feel of the open water. In the view of the interns who helped out at the release, it was a successful release because, they said, "No one cried." Apparently, some children become very attached to their adopted turtle sister on the short walk from the building to the bay. I can relate. I am attached, connected to those turtles, as surely as we are all connected to this blue-green paradise we are so fortunate to call home.

> Natural beauty is ubiquitous, but you have to meet it half way.
> Nature addresses our senses, but it takes a modicum of science
> to transform sensory experience into aesthetic sensibility.
> —J. BAIRD CALLICOTT, *Wetland Gloom and Wetland Glory*

After only a brief jaunt, an eager adventurer can become all too well acquainted with the sensory experience of a wetland, from the malodorous, foot-trapping mud to the too-tall tangles of cattails and shrubs. A different wetland path could yield easy walking on mossy paths, sky-blue water meandering at the edge, and light filtering through a cool shady canopy. Turning a corner can reveal a jaw-dropping view or an unexpected experience — a sora rail scuttling away, a baby least weasel calling a high-pitched peal. But it takes a longer study to even begin to conceive of the intricate connections and cycles that control the structure and function of a complex ecosystem. These revelations may only come from endless days of data collection in the lab or field, followed by many sedentary hours in front of a computer; here, in these sterile settings, the elegant patterns in nature become manifest.

In describing the forces of nature that drove him to spend a year on an isolated Cape Cod beach even though he had intended to leave after a fortnight, writer-naturalist Henry Beston wrote in *The Outermost House*:

> As the year lengthened into autumn, the beauty and mystery of this earth and outer sea so possessed and held me that I could not go. The world to-day is sick to its thin blood for lack of elemental things, for fire before the hands, for water welling from the earth, for air, for the dear earth itself underfoot.

At the end of his year, he reflected:

> Because I had known this outer and secret world, and been able to live as I had lived, reverence and gratitude greater and deeper than ever possessed me, sweeping every emotion else aside. . . . The ancient values of dignity, beauty and poetry which sustain it are of Nature's inspiration; they are born of the mystery and beauty of the world.

It is this mystery and beauty that motivated us to write this book. Together we two authors have worked, studied and laughed in bayous, backwaters, bogs, marshes, and meadows. Our work has left us deep in gratitude and inspired to reciprocity, to give back to the natural world that has so enriched us through our work and our explorations. Unraveling the complexities of a small fen or a thousand-acre bog, revealing the inextricably entwined relationships between the water flows, the natural chemistry, the soils, the flora and fauna, is our delight; understanding the overarching science of how the stagnant water becomes deprived of oxygen, setting off a chain of chemical reactions and evolutionary adaptations never ceases to amaze. Into this dark magic is added the changes wrought by humans, creating a mystery that saddens, but still tantalizes as a puzzle to be solved. Nature — whether beaver ponds or western mountains, southern seas or farmland skies — carries the poetry, and science reveals its importance. Once known, these links that attach us to the earth must be honored. Humans are ethically bound and evolutionarily inclined to care for all pieces of this planet, and we have faith that our hearts are big enough and our minds are sharp enough to rise to the tasks before us.

Acknowledgments

We would like to thank all those who shared their stories and to encourage others to tell their tales — the good, the bad, the ugly, the funny. It is in the sharing of stories that the true wonder, and the real plight, of wetlands is felt and understood. We also thank them for all they do to understand and protect our wetlands. In addition, we would like to thank Dr. John Harris for his guidance in turning real-world adventures into useful prose. We are also grateful to authors Sy Montgomery and Tom Wessels, as well as Nancy Monette and Ingeborg Hegemann, for reading early drafts of the book, offering supportive feedback, and writing short reviews to help us in the publication process. Finally, we want to thank Rob Koning, who inspired us to "write a book — isn't that what academics do?"; and Dietrich Earnhart, who encouraged us through the whole process.

References

Abbott Marshlands. n.d. Accessed December 10, 2014. http://abbottmarshlands.org/.

Abrams, Marc D. 1998. "The Red Maple Paradox." *BioScience* 48 (5): 355-"64.

Acreman, M., and J. Holden. 2013. "How Wetlands Affect Floods." *Wetlands* 33 (5): 773-86.

Adamowicz, Susan C., and Charles T. Roman. 2002. *Initial Ecosystem Response of Salt Marshes to Ditch Plugging and Pool Creation: Experiments at Rachel Carson National Wildlife Refuge (Maine).* Narragansett, RI: USGS Patuxent Wildlife Research Center, Costal Research Field Station.

Alber, Merryl, Erick M. Swenson, Susan C. Adamowicz, and Irving A. Mendelssohn. 2008. "Salt Marsh Dieback: An Overview of Recent Events in the US." *Estuarine, Coastal and Shelf Science* 80 (1): 1-11.

Allen, G. A., L. J. McCormick, J. R. Jantzen, K. L. Marr, and B. N. Brown. 2017. "Distributional and Morphological Differences between Native and Introduced Common Reed (*Phragmites australis*, Poaceae) in Western Canada." *Wetlands* 37 (5): 819-27.

Altieri, Andrew H., Mark D. Bertness, Tyler C. Coverdale, Eric E. Axelman, Nicholas C. Herrmann, and P. Lauren Szathmary. 2013. "Feedbacks Underlie the Resilience of Salt Marshes and Rapid Reversal of Consumer-Driven Die-Off." *Ecology* 94 (7): 1647-57.

Altieri, Andrew H., Mark D. Bertness, Tyler C. Coverdale, Nicholas C. Herrmann, and Christine Angelini. 2012. "A Trophic Cascade Triggers Collapse of a Salt-Marsh Ecosystem with Intensive Recreational Fishing." *Ecology* 93 (6): 1402-10.

Amon, James P., Carol A. Thompson, Quentin J. Carpenter, and James Miner. 2002.

"Temperate Zone Fens of the Glaciated Midwestern USA." *Wetlands* 22 (2): 301–17.

Artigas, Francisco, Jin Young Shin, Christine Hobble, Alejandro Marti-Donati, Karina V. R. Schäfer, and Ildiko Pechmann. 2015. "Long Term Carbon Storage Potential and CO_2 Sink Strength of a Restored Salt Marsh in New Jersey." *Agricultural and Forest Meteorology* 200:313–21.

Ashworth, Sharon M. 1997. "Comparison between Restored and Reference Sedge Meadow Wetlands in South-Central Wisconsin." *Wetlands* 17 (4): 518–27.

Associated Press. 2006. "Conn. Scientists Investigate Marsh Die-Off." *Washington Post*, June 26, 2006. http://www.washingtonpost.com/wp-dyn/content/article/2006/06/26/AR2006062601105.html.

Baldwin, Robert F., Aram J. K. Calhoun, and Phillip G. deMaynadier. 2006. "Conservation Planning for Amphibian Species with Complex Habitat Requirements: A Case Study Using Movements and Habitat Selection of the Wood Frog *Rana sylvatica*." *Journal of Herpetology* 40 (4): 442–53.

Barendregt, A., and C. W. Swarth. 2013. "Tidal Freshwater Wetlands: Variation and Changes." *Estuaries and Coasts* 36 (3): 445–56.

Beals, Whitney, and Peter Westover. 1971. *The Pine Creek and Mill River Watersheds, Fairfield, Connecticut: An Ecological Guide to Open Space Land Use*. Yale University School of Forestry.

Beck, Robert E. 1994. "The Movement in the United States to Restoration and Creation of Wetlands." *Natural Resources Journal* 34 (4): 781–822.

Bedford, Barbara L., and Kevin S. Godwin. 2003. "Fens of the United States: Distribution, Characteristics, and Scientific Connection versus Legal Isolation." *Wetlands* 23 (3): 608–29.

Berg, William E. 1992. "Large Mammals." In *The Patterned Peatlands of Minnesota*, edited by H. E. Wright Jr., B. Coffin, and N. Aaseng, 73–84. Minneapolis: University of Minnesota Press.

Bernard, J. M., and T. E. Lauve. 1995 "A Comparison of Growth and Nutrient Uptake in *Phalaris arundinacea L.* Growing in a Wetland and a Constructed Bed Receiving Landfill Leachate." *Wetlands* 15 (2): 176–82.

Bertness, Mark D., Caitlin P. Brisson, and Sinead M. Crotty. 2015. "Indirect Human Impacts Turn Off Reciprocal Feedbacks and Decrease Ecosystem Resilience." *Oecologia* 178 (1): 231–37.

Bertness, Mark D., and Tyler C. Coverdale. 2013. "An Invasive Species Facilitates the Recovery of Salt Marsh Ecosystems on Cape Cod." *Ecology* 94 (9): 1937–43.

Bertness, Mark D., Caitlin Crain, Christine Holdredge, and Nicholas Sala. 2008. "Eutrophication and Consumer Control of New England Salt Marsh Primary Productivity." *Conservation Biology* 22 (1): 131–39.

Bertness, Mark D., and Brian R. Silliman. 2008. "Consumer Control of Salt Marshes Driven by Human Disturbance." *Conservation Biology* 22 (3): 618–23.

Beston, H. 1956. *The Outermost House: A Year of Life on the Great Beach of Cape Cod.* New York: Viking.

Bledsoe, B. P., and T. H. Shear. 2000 "Vegetation along Hydrologic and Edaphic Gradients in a North Carolina Coastal Plain Creek Bottom and Implications for Restoration." *Wetlands* 20 (1): 126–47.

Bloomquist, Craig K., and C. Neilsen. 2010. "Demography of Unexploited Beavers in Southern Illinois." *Journal of Wildlife Management* 74 (2): 228–35.

Blossey, B. 2002. "Purple Loosestrife." In *Biological Control of Invasive Plants in the Eastern United States,* edited by Roy Van Driesche, B. Blossey, M. Hoddle, S. Lyon, and Richard Reardon, 149–50. Morgantown, WV: Forest Health Technology Enterprise Team.

Blossey, B., L. C. Skinner, and J. Taylor. 2001. "Impact and Management of Purple Loosestrife (*Lythrum salicaria*) in North America." *Biodiversity and Conservation* 10:1787–807.

Bongiorno, S. F., J. R. Trautman, T. J. Steinke, S. Kawa-Raymond, and D. Warner. 1984. "A Study of Restoration in Pine Creek Salt Marsh, Fairfield, Connecticut." In the *Proceedings of the 11th Annual Conference in Wetlands Restoration and Creation,* edited by Frederick J. Webb, Jr. Tampa, FL: Hillsborough Community College.

Boulton, A. J. 1989. "Over-summering Refuges of Aquatic Macroinvertebrates in Two Intermittent Streams in Central Victoria." *Transactions of the Royal Society of South Australia, Adelaide* 113 (1): 23–34.

Bragazza, Luca. 2008. "A Climatic Threshold Triggers the Die-Off of Peat Mosses during an Extreme Heat Wave." *Global Change Biology* 14 (11): 2688–95.

Bragina, Anastasia, Lisa Oberauner-Wappis, Christin Zachow, Bettina Halwachs, Gerhard G. Thallinger, Henry Müller, and Gabriele Berg. 2014. "The Sphagnum Microbiome Supports Bog Ecosystem Functioning under Extreme Conditions." *Molecular Ecology* 23 (18): 4498–510.

Brisson, Caitlin P., Tyler C. Coverdale, and Mark D. Bertness. 2014. "Salt Marsh Die-Off and Recovery Reveal Disparity between the Recovery of Ecosystem Structure and Service Provision." *Biological Conservation* 179:1–5.

Brooks, Robert T. 2005. "A Review of Basin Morphology and Pool Hydrology of Isolated Ponded Wetlands: Implications for Seasonal Forest Pools of the Northeastern United States." *Wetlands Ecology and Management* 13 (3): 335–48.

———. 2009. "Potential Impacts of Global Climate Change on the Hydrology and Ecology of Ephemeral Freshwater Systems of the Forests of the Northeastern United States." *Climatic Change* 95 (3–4): 469–83.

Brown, Stephen C., and Peter L. M. Veneman. 2001. "Effectiveness of Compensatory Wetland Mitigation in Massachusetts, USA." *Wetlands* 21 (4): 508–18.

Burchsted, Denise, Melinda Daniels, Robert Thorson, and Jason Vokoun. 2010. "The River Discontinuum: Applying Beaver Modifications to Baseline Conditions for Restoration of Forested Headwaters." *BioScience* 60 (11): 908–22.

Burgin, A. J., Julia G. Lazar, Peter M. Groffman, Arthur J. Gold, and D. Q. Kellogg. 2012. "Balancing Nitrogen Retention Ecosystem Services and Greenhouse Gas Disservices at the Landscape Scale." *Ecological Engineering* 56:26–35.

Burne, Matt. 2013. "An Unusual 'Moss.'" *Wicked Big Puddles* (blog). http://wickedbig puddles.blogspot.com/2013/08/an-unusual-moss.html.

Calhoun, Aram J. K., and Phillip G. deMaynadier, eds. 2007. *Science and Conservation of Vernal Pools in Northeastern North America: Ecology and Conservation of Seasonal Wetlands in Northeastern North America*. Boca Raton, FL: CRC Press.

Carter, Timothy C. 2006. "Indiana Bats in the Midwest: The Importance of Hydric Habitats." *Journal of Wildlife Management* 70 (5): 1185–90.

Chen, Hongjun. 2011. "Surface-Flow Constructed Treatment Wetlands for Pollutant Removal: Applications and Perspectives." *Wetlands* 31 (4): 805–14.

Cherry, J. A., and L. Gough. 2009. "Trade-Offs in Plant Responses to Herbivory Influence Trophic Routes of Production in a Freshwater Wetland." *Oecologia* 161 (3): 549–57. https://doi.org/10.1007/s00442-009-1408-8.

Chmura, Gail L. 2013. "What Do We Need to Assess the Sustainability of the Tidal Salt Marsh Carbon Sink?" *Ocean & Coastal Management* 83:25–31.

Colbert, Nathan K., Robert F. Baldwin, and Rachel K. Thiet. 2011. "A Developer-Initiated Conservation Plan for Pool-Breeding Amphibians in Maine, USA: A Case Study." *Journal of Conservation Planning* 7:27–38.

Colburn, Elizabeth A. 2004. *Vernal Pools: Natural History and Conservation*. Blacksburg: McDonald & Woodward.

Colburn, Elizabeth A., Stephen C. Weeks, and Sadie K. Reed. 2007. "Diversity and Ecology of Vernal Pool Invertebrates." In *Science and Conservation of Vernal Pools in Northeastern North America*, edited by Aram J. K. Calhoun and Phillip G. deMaynadier, 105–26. Boca Raton, FL: CRC Press.

Cole, Charles Andrew, and Robert P. Brooks. 2000. "Patterns of Wetland Hydrology in the Ridge and Valley Province, Pennsylvania, USA." *Wetlands* 20 (3): 438–47.

Cole, Charles Andrew, and Deborah Shafer. 2002. "Section 404 Wetland Mitigation and Permit Success Criteria in Pennsylvania, USA, 1986–1999." *Environmental Management* 30 (4): 508–15.

Connecticut Department of Energy and Environmental Protection. n.d. "Eastern Spadefoot Toad (*Scaphiopus holbrookii*)." Accessed January 8, 2016. http://www .ct.gov/deep/cwp/view.asp?a=2723&q=326002.

Cornell University Laboratory of Ornithology. 2016. "State of the Birds 2016." Accessed June 11, 2018. http://www.stateofthebirds.org/2016/overview/.

Costanza, Robert, Octavio Pérez-Maqueo, M. Luisa Martinez, Paul Sutton, Sharolyn J. Anderson, and Kenneth Mulder. 2008. "The Value of Coastal Wetlands for Hurricane Protection." *AMBIO: A Journal of the Human Environment* 37 (4): 241–48.

Council on Environmental Policy. 1978. *National Environmental Policy Act (NEPA) Compliance Guide*. 40 C.F.R. parts 1500–1508, § 1508.20.

Coverdale, Tyler C., Andrew H. Altieri, and Mark D. Bertness. 2012. "Belowground Herbivory Increases Vulnerability of New England Salt Marshes to Die-Off." *Ecology* 93 (9): 2085–94.

Coverdale, Tyler C., Mark D. Bertness, and Andrew H. Altieri. 2013. "Regional Ontogeny of New England Salt Marsh Die-Off." *Conservation Biology* 27 (5): 1041–48.

Crain, C. M., and M. D. Bertness. 2005. "Community Impacts of a Tussock Sedge: Is Ecosystem Engineering Important in Benign Habitats?" *Ecology* 86:2695–704.

Cresswell, James E. 1991. "Capture Rates and Composition of Insect Prey of the Pitcher Plant *Sarracenia purpurea*." *American Midland Naturalist* 125 (1): 1–9.

Cronk, Julie K., and M. Siobhan Fennessy. 2001. *Wetland Plants: Biology and Ecology.* Boca Raton, FL: Lewis Publishers.

Cronon, William. 2003. *Changes in the Land: Indians, Colonists, and the Ecology of New England,* rev. ed. New York: Hill and Wang.

Crum, Howard. 1992. *A Focus on Peatlands and Peat Mosses.* Ann Arbor: University of Michigan Press.

Curtis, Linda W. 2014. "The Last Sleep-Out." *Curtis to the Third Productions* (blog). http://curtistothethird.com/wp1/?page_id=496.

———. 2016. "What Good Are Sedges?" *Curtis to the Third Productions* (blog). February 13. http://curtistothethird.com/wp1/?p=1868.

Dahl, Thomas E. 2006. *Status and Trends of Wetlands in the Conterminous United States 1998 to 2004.* Washington, DC: US Fish and Wildlife Service.

———. 2011. *Status and Trends of Wetlands in the Conterminous United States 2004 to 2009.* Washington, DC: US Fish and Wildlife Service.

Davis, Whit. 2008. "The Wisdom of 'Whit' Davis." Interview by Ken Simon. Working the Land. SimonPure Productions. http://www.workingtheland.com/interview-davis.htm.

Deegan, Linda A., David Samuel Johnson, R. Scott Warren, Bruce J. Peterson, John W. Fleeger, Sergio Fagherazzi, and Wilfred M. Wollheim. 2012. "Coastal Eutrophication as a Driver of Salt Marsh Loss." *Nature* 490 (7420): 388–92.

Denton, Robert D., and Stephen C. Richter. 2013. "Amphibian Communities in Natural and Constructed Ridge Top Wetlands with Implications for Wetland Construction." *Journal of Wildlife Management* 77 (5): 886–96.

Dillard, Annie. 1974. *Pilgrim at Tinker Creek.* New York: Harper's Magazine Press.

Dixon, Mark D. 2003. "Effects of Flow Pattern on Riparian Seedling Recruitment on Sandbars in the Wisconsin River, Wisconsin, USA." *Wetlands* 23 (1): 125–39.

Donahue, Brian. 2004. *The Great Meadow: Farmers and the Land in Colonial Concord.* New Haven, CT: Yale University Press.

Donner, Deahn M., Christine A. Ribic, Albert J. Beck, Dale Higgins, Dan Eklund, and Susan Reinecke. 2015. "Woodland Pond Salamander Abundance in Relation to Forest Management and Environmental Conditions in Northern Wisconsin." *Journal of North American Herpetology* 2015 (1): 34–42.

Duffy, J. Emmett. 2002. "Biodiversity and Ecosystem Function: The Consumer Connection." *Oikos* 99 (2): 201–19.

DuRant, Sarah E., and William A. Hopkins. 2008. "Amphibian Predation on Larval Mosquitoes." *Canadian Journal of Zoology* 86 (10): 1159–64.

Eastman, John. 1995. *The Book of Swamp and Bog*. Pennsylvania: Stackpole Books.

Egan, Dave. 1990. "Historic Initiatives in Ecological Restoration." *Ecological Restoration* 8 (2): 83–90.

Ehrlich, Paul, David S. Dobkin, and Darryl Wheye. 1988. *Birder's Handbook*. New York: Simon and Schuster.

Elsey-Quirk, T., and M. A. Leck. 2015. "Patterns of Seed Bank and Vegetation Diversity along a Tidal Freshwater River." *American Journal of Botany* 102 (12): 1996–2012. https://doi.org/10.3732/ajb.1500314.

Errington, Paul L. 1957. *Of Men and Marshes*. New York: MacMillan.

Farnsworth, E. J., and D. Ellis. 2001. "Is Purple Loosestrife (*Lythrum salicaria*) an Invasive Threat to Freshwater Wetlands? Conflicting Evidence from Several Ecological Metrics." *Wetlands* 21 (2): 199–209.

Galatowitsch, Susan M., Neil O. Anderson, and Peter D. Ascher. 1999. "Invasiveness in Wetland Plants in Temperate North America." *Wetlands* 19 (4): 733–55.

Garbisch, Edgar W. 2005. "Hambleton Island Restoration: Environmental Concern's First Wetland Creation Project." *Ecological Engineering* 24 (4): 289–307.

Gedan, Keryn B., Andrew H. Altieri, and Mark D. Bertness. 2011. "Uncertain Future of New England Salt Marshes." *Marine Ecology Progress Series* 434:229–37.

Gibbs, James P., and Alvin R. Breisch. 2001. "Climate Warming and Calling Phenology of Frogs near Ithaca, New York, 1900–1999." *Conservation Biology* 15 (4): 1175–78.

Gobster, P. H., J. I. Nassauer, T. C. Daniel, and G. Fry. 2007. "The Shared Landscape: What Does Aesthetics Have to Do with Ecology?" *Landscape Ecology* 22 (7): 959–72.

Godwin, Kevin S., James P. Shallenberger, Donald J. Leopold, and Barbara L. Bedford. 2002. "Linking Landscape Properties to Local Hydrogeologic Gradients and Plants Species Occurrence in Minerotrophic Fens of New York State, USA: A Hydrogeologic Setting (HGS)." *Wetlands* 22 (4): 722–37.

Goodby, Robert G., Paul Bock, Edward Bouras, Christopher Dorion, A. Garrett Evans, Tonya Largy, Stephen Pollock, Heather Rockwell, and Arthur Spiess. 2014. "The Tenant Swamp Site and Paleoindian Domestic Space in Keene, New Hampshire." *Archaeology of Eastern North America* 42:129–64.

Grevstad, F. S. 2006. "Ten-Year Impacts of the Biological Control Agents *Galerucella pusilla* and *G. calmariensis* (Coleoptera: Chrysomelidae) on Purple Loosestrife (*Lythrum salicaria*) in Central New York State." *Biological Control* 39:1–8.

Groc, Isabelle. 2010. "Beavers Sign Up to Fight Effects of Climate Change." *Discover*, April 19, 2010. http://discovermagazine.com/2010/apr/19-beavers-sign-up-fight-effects-climate-change.

Guthery, Fred S., and Fred C. Bryant. 1982. "Status of Playas in the Southern Great Plains." *Wildlife Society Bulletin* 10 (4): 309–17.

Hager, Heather A., and Rolf D. Vinebrooke. 2004. "Positive Relationships between Invasive Purple Loosestrife (*Lythrum salicaria*) and Plant Species Diversity and Abundance in Minnesota Wetlands." *Canadian Journal of Botany* 82 (6): 763–73.

Hardisky, Tom. 2011. *Beaver Management in Pennsylvania (2010–2019)*. Harrisburg, PA: Pennsylvania Game Commission. https://fyi.uwex.edu/beaver/files/2011/10/Pennsylvania-Beaver-Mgt-Plan-2010.pdf.

Harms, Tyler M., and Stephen J. Dinsmore. 2013. "Habitat Associations of Secretive Marsh Birds in Iowa." *Wetlands* 33 (3): 561–71.

Hauser, S., M. S. Meixler, and M. Laba. 2015. "Quantification of Impacts and Ecosystem Services Loss in New Jersey Coastal Wetlands Due to Hurricane Sandy Storm Surge." *Wetlands* 35 (6): 1137–48.

Healy, M. T., and J. B. Zedler. 2010. "Setbacks in Replacing *Phalaris arundinacea* Monotypes with Sedge Meadow Vegetation." *Restoration Ecology* 18 (2): 155–64.

Hearne, Samuel, and Joseph Burr Tyrrell. 1911. *A Journey from Prince of Wales' Fort in Hudson Bay to the Northern Ocean, in the Years 1769, 1770, 1771, and 1772*. Toronto: Champlain Society.

Heter, Elmo W. 1950. "Transplanting Beavers by Airplane and Parachute." *Journal of Wildlife Management* 14 (2): 143–47.

Hewitt, Nina, and Kiyoko Miyanishi. 1997. "The Role of Mammals in Maintaining Plant Species Richness in a Floating Typha Marsh in Southern Ontario." *Biodiversity and Conservation* 6 (8): 1085–1102.

Hinz, Hariet L., Mark Schwarzländer, André Gassmann, and Robert S. Bourchier. 2014. "Successes We May Not Have Had: A Retrospective Analysis of Selected Weed Biological Control Agents in the United States." *Invasive Plant Science and Management* 7 (4): 565–79.

Hipp, Andrew L., Paul E. Rothrock, and Eric H. Roalson. 2009. "The Evolution of Chromosome Arrangements in *Carex* (Cyperaceae)." *Botanical Review* 75 (1): 96–109.

Historica Canada. n.d. "Ducks Unlimited Canada." Accessed Dec. 10, 2015. http://www.thecanadianencyclopedia.ca/en/article/ducks-unlimited-canada.

Hoffmann, Carl Christian, C. Kjaergaard, J. Uusi-Kämppä, H. C. Hansen, and B. Kronvang. 2009. "Phosphorus Retention in Riparian Buffers: Review of Their Efficiency." *Journal of Environmental Quality* 38 (5): 1942–55.

Holdredge, Christine, Mark D. Bertness, and Andrew H. Altieri. 2009. "Role of Crab Herbivory in Die-Off of New England Salt Marshes." *Conservation Biology* 23 (3): 672–79.

Hood, Glynnis A. 2011. *The Beaver Manifesto*. British Columbia: Rocky Mountain Books.

Hood, Glynnis A., and Suzanne E. Bayley. 2008. "Beaver (*Castor canadensis*) Mitigate the Effects of Climate on the Area of Open Water in Boreal Wetlands in Western Canada." *Biological Conservation* 141 (2): 556–67.

Hood, W. Gregory. 2012. "Beaver in Tidal Marshes: Dam Effects on Low-Tide Channel Pools and Fish Use of Estuarine Habitat." *Wetlands* 32 (3): 401–10.

Hossler, Katie, Virginie Bouchard, M. Siobhan Fennessy, Serita D. Frey, Evelyn Anemaet, and Ellen Herbert. 2011. "No-Net-Loss Not Met for Nutrient Function in Freshwater Marshes: Recommendations for Wetland Mitigation Policies." *Ecosphere* 2 (7): 1–36.

Hough, Palmer, and Morgan Robertson. 2009. "Mitigation under Section 404 of the Clean Water Act: Where It Comes From, What It Means." *Wetlands Ecology and Management* 17 (1): 15–33.

Hunter, Malcolm L., John Albright, and Jane Arbuckle. 1992. *The Amphibians and Reptiles of Maine.* Orono: University of Maine Press.

Jakubowski, Andrew R., Michael D. Casler, and Randall D. Jackson. 2012. "Genetic Evidence Suggests a Widespread Distribution of Native North American Populations of Reed Canarygrass." *Biological Invasions* 15 (2): 261–68.

Janowiak, Maria K., Louis R. Iverson, David J. Mladenoff, Emily Peters, Kirk R. Wythers, Weimin Xi, Leslie A. Brandt, Patricia R. Butler, et al. 2014. *Forest Ecosystem Vulnerability Assessment and Synthesis for Northern Wisconsin and Western Upper Michigan: A Report from the Northwoods Climate Change Response Framework Project.* Gen. Tech. Rep. NRS-136. Newtown Square, PA: US Department of Agriculture, Forest Service, Northern Research Station.

Jarecke, Karla M., Terrance D. Loecke, and Amy J. Burgin. 2016. "Coupled Soil Oxygen and Greenhouse Gas Dynamics under Variable Hydrology." *Soil Biology and Biochemistry* 95:164–72.

Jordan, William R., and George M. Lubick. 2011. *Making Nature Whole: A History of Ecological Restoration.* Washington, DC: Island Press.

Joyal, Lisa A., Mark McCollough, and Malcolm L. Hunter. 2001. "Landscape Ecology Approaches to Wetland Species Conservation: A Case Study of Two Turtle Species in Southern Maine." *Conservation Biology* 15 (6): 1755–62.

Kart, J., R. Regan, S. R. Darling, C. Alexander, K. Cox, M. Ferguson, S. Parren, K. Royar, and B. Popp, eds. 2005. *Vermont's Wildlife Action Plan.* Waterbury: Vermont Fish & Wildlife Department. http://vtfishandwildlife.com/sites/fish andwildlife/files/documents/About%20Us/Budget%20and%20Planning/VT _Willdife_Action_Plan_Main_Document.pdf.

Kerney, Ryan, Eunsoo Kim, Roger P. Hangarter, Aaron A. Heiss, Cory D. Bishop, and Brian K. Hall. 2011. "Intracellular Invasion of Green Algae in a Salamander Host." *Proceedings of the National Academy of Sciences* 108 (16): 6497–502.

Kimmelman, M. 2012. "River of Hope in the Bronx." *New York Times*, July 19. https:// www.nytimes.com/2012/07/22/arts/design/bronx-river-now-flows-by-parks .html.

Kiviat, Erik. 1978. "Vertebrate Use of Muskrat Lodges and Burrows." *Estuaries and Coasts* 1 (3): 196–200.

Klubertanz, Dale. 2016. *The Story of Friends of Pheasant Branch Conservancy: 20 Years of Advocacy.* Vimeo. Video, 20:37. https://vimeo.com/121601915.

Kopp, Robert E., Benjamin P. Horton, Andrew C. Kemp, and Claudia Tebaldi. 2015. "Past and Future Sea-Level Rise along the Coast of North Carolina, USA." *Climatic Change* 132 (4): 693–707.

Kost, Michael A., Dennis A. Albert, Joshua G. Cohen, Bradford S. Slaughter, Rebecca K. Schillo, Christopher R. Weber, and Kim A. Chapman. 2007. "Natural Communities of Michigan: Classification and Description." *Michigan Natural Features Inventory* Report Number 2007-21, Lansing, MI.

Ksander, Yaël. 2008. "The Hundred Year Flood." Indiana Public Media. http://indiana publicmedia.org/momentofindianahistory/100-year-flood/.

Laidig, Kim J., and Robert A. Zampella. 1999. "Community Attributes of Atlantic White Cedar (*Chamaecyparis thyoides*) Swamps in Disturbed and Undisturbed Pinelands Watersheds." *Wetlands* 19 (1): 35–49.

LaRoe, Edward T. 1978. "Mitigation: A Concept for Wetland Restoration." In *Proceedings of the National Wetland Protection Symposium, June 6–8, 1977, Reston, Virginia,* 221–24. Reston, VA: US Fish and Wildlife Service.

Larson, J. S., and J. A. Kusler. 1979. Preface to *Wetland Functions and Values: The State of Our Understanding; Proceedings of the National Symposium on Wetlands,* edited by P. E. Greeson, J. R. Clark, and J. E. Clark. Minneapolis: American Water Resources Association.

Lavergne, S., and J. Molofsky. 2004. "Reed Canary Grass (*Phalaris arundinacea*) as a Biological Model in the Study of Plant Invasions." *Critical Reviews in Plant Sciences* 23 (5): 415–29.

Lavoie, C. 2010. "Should We Care about Purple Loosestrife? The History of an Invasive Plant in North America." *Biological Invasions* 12:1967–99.

Lawrence, Beth A., and Joy B. Zedler. 2011. "Formation of Tussocks by Sedges: Effects of Hydroperiod and Nutrients." *Ecological Applications* 21 (5): 1745–59.

Leahy, C. W., B. Cassie, and R. K. Walton, eds. 2006. Massachusetts Butterfly Atlas 1986–1990. Mass Audubon. https://www.massaudubon.org/butterflyatlas.

Leck, Mary Allessio. 2004. "Seeds, Seed Banks and Wetlands." *Seed Science Research* 14 (3): 259–66.

Lesser, Christopher R. 2007. *Open Marsh Water Management: A Source Reduction Technique for Mosquito Control.* Delaware Mosquito Control Section. http://dnrec .delaware.gov/fw/mosquito/Documents/OMWM%20Article%2011.05.07.pdf.

Lewis, Richard C. 2007. "Cape Salt Marsh Decline Linked to Native Crab." *Boston Globe,* November 19. http://archive.boston.com/news/science/articles/2007 /11/19/cape_salt_marsh_decline_linked_to_native_crab/.

Lindgren, C. J., and D. Walker. 2012. "Growth Rate, Seed Production, and Assessing the Spatial Risk of *Lythrum salicaria* Using Growing Degree-Days." *Wetlands* 32 (5): 885–93.

Little, Amanda, Glenn R. Guntenspergen, and Timothy F. H. Allen. 2012. "Wetland Vegetation Dynamics in Response to Beaver (*Castor canadensis*) Activity at Multiple Scales." *Ecoscience* 19 (3): 246–57.

Lobell, Jarrett A., and Samir S. Patel. 2010. "Bog Bodies Rediscovered: True Tales from the Peat Marshes of Northern Europe." *Archaeology* 63 (3): 22–26.

Lopez, Barry, and Debra Gwartney, eds. 2006. *Home Ground: Language for an American Landscape*, 1st ed. San Antonio, TX: Trinity University Press.

Lotts, Kelly, and Thomas Naberhaus, coordinators. 2014. "Mullberry Wing: *Poanes massasoit* (Scudder, 1864)." Butterflies and Moths of North America. Accessed January 2015. http://www.butterfliesandmoths.org/species/Poanes-massasoit.

Lovett, Jennifer. 2016. *Beavers Away!* Edina, MN: Beaver's Pond Press.

Madison, Dale M. 1997. "The Emigration of Radio-Implanted Spotted Salamanders, Ambystoma Maculatum." *Journal of Herpetology* 31 (4): 542.

Magee, Teresa K., and Mary E. Kentula. 2005. "Response of Wetland Plant Species to Hydrologic Conditions." *Wetlands Ecology and Management* 13 (2): 163–81.

Maine Department of Transportation and Reed & D'Andrea. 1974. *Saltmarsh Relocation Restoration in Maine*. Maine State Library: Transportation Documents. Paper 37. https://digitalmaine.com/mdot_docs/37/.

Mal, T. K., J. Lovett-Doust, L. Lovett-Doust, and G. A. Mulligan. 1992. "The Biology of Canadian Weeds: 100. *Lythrum salicaria*." *Canadian Journal of Plant Science* 72: 1305–30.

Martin, Craig E., and Sarah K. Francke. 2015. "Root Aeration Function of Baldcypress Knees (*Taxodium distichum*)." *International Journal of Plant Sciences* 176 (2): 170–73.

Martin, Elizabeth, Julie Duke, Mathew Pelkki, Edgar C. Clausen, and Danielle Julie Carrier. 2010. "Sweetgum (*Liquidambar styraciflua L.*): Extraction of Shikimic Acid Coupled to Eilute Acid Pretreatment." *Applied Biochemistry and Biotechnology* 162 (6): 1660–68.

Mashantucket Pequot Museum & Research Center. n.d. "Battlefield Sites: Pequot War Battlefield Sites." Battlefields of the Pequot War. Accessed June 7, 2016. http://pequotwar.org/about/sites/.

Massachusetts Division of Fisheries and Wildlife, Natural Heritage and Endangered Species Program. 2015. "Eastern Spadefoot." Massachusetts Division of Fisheries and Wildlife. https://www.mass.gov/files/documents/2016/08/no/scaphiopus-holbrookii.pdf.

———. 2016. "Blue-Spotted Salamander." Massachusetts Division of Fisheries and Wildlife. https://www.mass.gov/files/documents/2017/01/qd/ambystoma-laterale.pdf.

Maurer, Deborah A., and Joy B. Zedler. 2002. "Differential Invasion of a Wetland

Grass Explained by Tests of Nutrients and Light Availability on Establishment and Clonal Growth." *Oecologia* 131 (2): 279–88.

McCall, Thomas C., Thomas P. Hodgman, Duane R. Diefenbach, and Ray B. Owen. 1996. "Beaver Populations and Their Relation to Wetland Habitat and Breeding Waterfowl in Maine." *Wetlands* 16 (2): 163–72.

McCormack, L. n.d. "Species Spotlight: Eastern Spadefoot Toad." Accessed January 8, 2016. http://www.albanypinebush.org/pdf/SpadefootToad.pdf.

McDonald, John E. Jr., and Todd K. Fuller. 2005. "Effects of Spring Acorn Availability on Black Bear Diet, Milk Composition, and Cub Survival." *Journal of Mammalogy* 86 (5): 1022–28.

McIninch, S. M., and D. R. Biggs. 1993. "Mechanisms of Tolerance to Saturation of Selected Woody Plants." *Wetland Journal Environmental Concern* 2:25–27.

McIninch, S., E. Garbisch, and D. R. Biggs. 1994. "The Benefits of Wet-Acclimating Woody Wetland Plant Species." *Wetland Journal Environmental Concern* 6:19–23.

McLean, Stuart. 2008. "Bodies from the Bog: Metamorphosis, Non-human Agency and the Making of 'Collective' Memory." *Trames Journal of the Humanities and Social Sciences* 12 (3): 299–308.

Mcleod, Elizabeth, Gail L. Chmura, Steven Bouillon, Rodney Salm, Mats Björk, Carlos M. Duarte, Catherine E. Lovelock, William H. Schlesinger, and Brian R. Silliman. 2011. "A Blueprint for Blue Carbon: Toward an Improved Understanding of the Role of Vegetated Coastal Habitats in Sequestering CO_2." *Frontiers in Ecology and the Environment* 9 (10): 552–60.

McLoughlin, Philip D., Jesse S. Dunford, and Stan Boutin. 2005. "Relating Predation Mortality to Broad-Scale Habitat Selection." *Journal of Animal Ecology* 74 (4): 701–7.

McMaster, Robert T., and Nancy D. McMaster. 2001. "Composition, Structure, and Dynamics of Vegetation in Fifteen Beaver-Impacted Wetlands in Western Massachusetts." *Rhodora* 103 (915): 293–320.

McQueen, Cyrus B. 1990. *Field Guide to the Peat Mosses of Boreal North America.* Lebanon, NH: University Press of New England.

Mitsch, W. J., John W. Day, Li Zhang, and Robert R. Lane. 2005. "Nitrate-Nitrogen Retention in Wetlands in the Mississippi River Basin." *Ecological Engineering* 24 (4): 267–78.

Mitsch, W. J., and J. G. Gosselink. 2000. *Wetlands.* New York: John Wiley & Sons.

Moerman, Daniel E. 1998. *Native American Ethnobotany.* Portland, OR: Timber Press.

Moore, Peter D. 1999. "Sprucing Up Beaver Meadows." *Nature* 400 (6745): 622–23.

Moreno-Mateos, David, Mary E. Power, Francisco A. Comín, and Roxana Yockteng. 2012. "Structural and Functional Loss in Restored Wetland Ecosystems." *PLOS Biology* 10 (1): e1001247.

Morrison, J. A. 2002. "Wetland Vegetation before and after Experimental Purple Loosestrife Removal." *Wetlands* 22 (1): 159–69.

Müller-Schwarze, Dietland. 2011. *The Beaver: Its Life and Impact*. Ithaca, NY: Cornell University Press.

Müller-Schwarze, Dietland, and Lixing Sun. 2003. *The Beaver: Natural History of a Wetlands Engineer*. Ithaca, NY: Cornell University Press.

Naiman, Robert J., Carol A. Johnston, and James C. Kelley. 1988. "Alteration of North American Streams by Beaver." *BioScience* 38 (11): 753–62.

National Research Council, Committee on Mitigating Wetland Losses. 2001. *Compensating for Wetland Losses under the Clean Water Act*. National Academy of Sciences. Washington DC: National Academies Press.

National Weather Service. n.d. Wabash River at Terra Haute. Advanced Hydrologic Prediction Service. Accessed July 20, 2015. https://water.weather.gov/ahps2/hydrograph.php?gage=hufi3&wfo=ind.

Nekola, Jeffrey C. 1994. "The Environment and Vascular Flora of Northeastern Iowa Fen Communities." *Rhodora* 96 (886): 121–69.

Nellemann, Christian, Emily Corcoran, C. M. Duarte, L. Valdés, C. De Young, L. Fonseca, and G. Grimsditch, eds. 2009. "Blue Carbon: A Rapid Response Assessment." Arendal, Norway: United Nations Environment Programme, GRID-Arendal.

New England Estuarine Research Society. n.d. New England Wetland Dieback. Accessed April 27, 2015. http://www.neers.org.

New England Wildflower Society. n.d. Go Botany. Accessed July 8, 2016. https://gobotany.newenglandwild.org/.

Nyffeler, Martin, and Bradley J. Pusey. 2014. "Fish Predation by Semi-aquatic Spiders: A Global Pattern." *PLOS One* 9 (6): e99459.

O'Connell, Timothy J., Robert P. Brooks, Diann J. Prosser, Mary T. Gaudette, Joseph P. Gyekis, Kimberly C. Farrell, and Mary Jo Casalena. 2013. "Wetland-Riparian Birds of the Mid-Atlantic Region." In *Mid-Atlantic Freshwater Wetlands: Advances in Wetlands Science, Management, Policy, and Practice*, edited by Robert P. Brooks and Denice Heller Wardrop, 269–311. New York: Springer.

Owen, Catherine R., Quentin J. Carpenter, and Calvin B. DeWitt. 1989. *Evaluation of Three Wetland Restorations Associated with Highway Projects*. Madison: Institute for Environmental Studies, University of Wisconsin.

Parker, V. T., and M. A. Leck. 1985. "Relationships of Seed Banks to Plant Distribution Patterns in a Freshwater Tidal Wetland." *American Journal of Botany* 72 (2): 161–74.

Paton, Peter W. C. 2005. "A Review of Vertebrate Community Composition in Seasonal Forest Pools of the Northeastern United States." *Wetlands Ecology and Management* 13 (3): 235–46.

Paton, Peter W. C., and William B. Crouch. 2002. "Using the Phenology of Pond-Breeding Amphibians to Develop Conservation Strategies." *Conservation Biology* 16 (1): 194–204.

Peach, Michelle, and Joy B. Zedler. 2006. "How Tussocks Structure Sedge Meadow Vegetation." *Wetlands* 26 (2): 322–35.

Pendleton, Linwood, Daniel C. Donato, Brian C. Murray, Stephen Crooks, W. Aaron Jenkins, Samantha Sifleet, Christopher Craft, James W. Fourqurean, J. Boone Kauffman, Núria Marbà, Patrick Megonigal, Emily Pidgeon, Dorothee Herr, David Gordon, and Alexis Baldera. 2012. "Estimating Global 'Blue Carbon' Emissions from Conversion and Degradation of Vegetated Coastal Ecosystems." *PLOS One* 7 (9): e43542.

Perry, James E., and Robert B. Atkinson. 1997. "Plant Diversity along a Salinity Gradient of Four Marshes on the York and Pamunkey Rivers in Virginia." *Castanea* 62 (2): 112–18.

Perry, James E., Donna Marie Bilkovic, Kirk J. Havens, and Carl Hershner. 2009. "Tidal Freshwater Marshes of the Mid-Atlantic and Southeastern United States." In *Tidal Freshwater Wetlands*, edited by A. Barendregt, D. Whigham, and A. Baldwin, 157–66. Weikersheim: Backhuys Publishers.

Petit, Lisa J. 1999. "Prothonotary Warbler (*Protonotaria citrea*)." The Birds of North America Online. Edited by A. Poole. Ithaca, NY: Cornell Lab of Ornithology. http://bna.birds.cornell.edu/bna/species/408.

Petranka, James W., Andrea W. Rushlow, and Mark E. Hopey. 1998. "Predation by Tadpoles of *Rana sylvatica* on Embryos of *Ambystoma maculatum*: Implications of Ecological Role Reversals by *Rana* (Predator) and *Ambystoma* (Prey)." *Herpetologica* 54 (1): 1–13.

Phelps, Quinton E., Sara J. Tripp, David P. Herzog, and James E. Garvey. 2015. "Temporary Connectivity: The Relative Benefits of Large River Floodplain Inundation in the Lower Mississippi River." *Restoration Ecology* 23 (1): 53–56.

Pinsky, Malin L., Greg Guannel, and Katie K. Arkema. 2013. "Quantifying Wave Attenuation to Inform Coastal Habitat Conservation." *Ecosphere* 4 (8): 1–16.

Pollock, Michael M., Morgan Heim, and Danielle Werner. 2003. "Hydrologic and Geomorphic Effects of Beaver Dams and their Influence on Fishes." *American Fisheries Society Symposium* 37:213–33.

Proceedings of the First Annual Conference on Restoration of Coastal Vegetation in Florida. 1974. Tampa, FL: Hillsborough Community College.

Raabe, Ellen A., and Richard P. Stumpf. 2016. "Expansion of Tidal Marsh in Response to Sea-Level Rise: Gulf Coast of Florida, USA." *Estuaries and Coasts* 39 (1): 145–57.

Rey, Jorge R., William E. Walton, Roger J. Wolfe, C. Roxanne Connelly, Sheila M. O'Connell, Joe Berg, Gabrielle E. Sakolsky-Hoopes, and Aimlee D. Laderman. 2012. "North American Wetlands and Mosquito Control." *International Journal of Environmental Research and Public Health* 9 (12): 4537–605.

Rheinhardt, Richard D. 2007. "Hydrogeomorphic and Compositional Variation among Red Maple (*Acer rubrum*) Wetlands in Southeastern Massachusetts." *Northeastern Naturalist* 14 (4): 589–604.

Rice, Graham. 2012. "The Flowering of *Symplocarpus*." *Plantsman*, n.s., 11 (1): 54–57.

Riffell, Samuel, Thomas Burton, and Margaret Murphy. 2006. "Birds in Depressional Forested Wetlands: Area and Habitat Requirements and Model Uncertainty." *Wetlands* 26 (1): 107–18.

Rittenhouse, Tracy A. G., and Raymond D. Semlitsch. 2007. "Distribution of Amphibians in Terrestrial Habitat Surrounding Wetlands." *Wetlands* 27 (1): 153–61.

Robb, James T. 2002. "Assessing Wetland Compensatory Mitigation Sites to Aid in Establishing Mitigation Ratios." *Wetlands* 22 (2): 435–40.

Roman, Charles T., William A. Niering, and R. Scott Warren. 1984. "Salt Marsh Vegetation Change in Response to Tidal Restriction." *Environmental Management* 8 (2): 141–49.

Rosell, Frank, Orsolya Bozser, Peter Collen, and Howard Parker. 2005. "Ecological Impact of Beavers *Castor fiber* and *Castor canadensis* and Their Ability to Modify Ecosystems." *Mammal Review* 35 (3–4): 248–76.

Rowinski, Christine. 1995. "Functions and Values of Forested/Scrub-Shrub Wetlands: Research Summary." New Hampshire Coastal Program. http://www.gpo.gov /fdsys/pkg/CZIC-qk174-r69-1995/html/CZIC-qk174-r69-1995.htm.

Rozsa, Ron. 1995. "Human Impacts on Tidal Wetlands: History and Regulations." In *Bulletin No. 34: Tidal Marshes of Long Island Sound: Ecology, History, and Restoration*. New London: Connecticut College Arboretum.

Rozsa, Ron, and Richard A. Orson. 1993. "Restoration of Degraded Salt Marshes in Connecticut." In *Proceedings of the 20th Annual Conference on Wetlands Restoration and Creation*, edited by Frederick Webb Jr., 196–205. Tampa, FL: Hillsborough Community College.

Rubbo, Michael J., Jessie L. Lanterman, Richard C. Falco, and Thomas J. Daniels. 2011. "The Influence of Amphibians on Mosquitoes in Seasonal Pools: Can Wetlands Protection Help to Minimize Disease Risk?" *Wetlands* 31 (4): 799–804.

Saltonstall, Kristin. 2002. "Cryptic Invasion by a Non-native Genotype of the Common Reed, *Phragmites australis*, into North America." *Proceedings of the National Academy of Sciences* 99 (4): 2445–49.

Sauer, J. R., D. K. Niven, J. E. Hines, D. J. Ziolkowski Jr., K. L. Pardieck, J. E. Fallon, and W. A. Link. 2017. *The North American Breeding Bird Survey, Results and Analysis 1966–2015*. Version 2.07.2017. Laurel, MD: USGS Patuxent Wildlife Research Center. https://www.mbr-pwrc.usgs.gov/bbs/.

Scanga, Sara E. 2011. "Effects of Light Intensity and Groundwater Level on the Growth of a Globally Rare Fen Plant." *Wetlands* 31 (4): 773–81.

Schenck, Elizabeth H. 1889. *The History of Fairfield, Fairfield County, Connecticut, from the Settlement of the Town in 1639 to 1818, Vol. I*. New York: published by the author.

Schlüter, Urte, and Robert M. Crawford. 2003. "Metabolic Adaptation to Prolonged Anoxia in Leaves of American Cranberry (*Vaccinium macrocarpon*)." *Physiologia Plantarum* 117 (4): 492–99.

Schooler, S. S., P. B. McEvoy, and E. M. Coombs. 2006. "Negative Per Capita Effects of Purple Loosestrife and Reed Canary Grass on Plant Diversity of Wetland Communities." *Diversity and Distributions* 12:351–63.

Seymour, R. S. 2004. "Dynamics and Precision of Thermoregulatory Responses of Eastern Skunk Cabbage *Symplocarpus foetidus.*" *Plant, Cell & Environment* 27 (8): 1014–22. https://doi.org/10.1111/j.1365-3040.2004.01206.x.

Shapiro, Arthur M. 1970. "The Biology of *Poanes viator* (Hesperiidae) with the Description of a New Subspecies." *Journal of Research on the Lepidoptera* 9 (2): 109–23.

Shuey, J. A. "Phylogeny and Biogeography of *Euphyes* Scudder (Hesperiidae)." 1993. *Journal of the Lepidopterists Society* 47 (4): 261–78.

Silliman, Brian Reed, and Mark D. Bertness. 2002. "A Trophic Cascade Regulates Salt Marsh Primary Production." *Proceedings of the National Academy of Sciences* 99 (16): 10500–505.

Small, Deborah A., George Loewenstein, and Paul Slovic. 2007. "Sympathy and Callousness: The Impact of Deliberative Thought on Donations to Identifiable and Statistical Victims." *Organizational Behavior and Human Decision Processes* 102 (2): 143–53.

Smith, Douglas G. 1992. "A New Freshwater Moss Animal in the genus *Plumatella* (Ectoprocta: Phylactolaemata: Plumatellidae) from New England (USA)." *Canadian Journal of Zoology* 70 (11): 2192–201.

Smith, Stephen M. 2006. *Report on Salt Marsh Dieback on Cape Cod.* North Truro, MA: National Park Service.

Snyder, S. A. 1993. "Ondatra zibethicus." Fire Effects Information System. https://www.fs.fed.us/database/feis/animals/mammal/onzi/all.html.

Society of Ecological Restoration International Science and Policy Working Group. 2004. *The SER International Primer on Ecological Restoration.* Tucson, AZ: Society for Ecological Restoration International.

Sperduto, D., and B. Kimball. 2011. *The Nature of New Hampshire: Natural Communities of the Granite State.* Durham, NH: University Press of New England.

Spurr, Katherine C. 2003. "Use of Geographic Information Systems and Global Positioning Technology to Map and Study Nesting Trends and Density Dynamics of a Heronry on the Upper Mississippi." Graduate thesis, St. Mary's University. http://www.gis.smumn.edu/GradProjects/SpurrK.pdf.

Stapanian, Martin A., Jean V. Adams, and Brian Gara. 2013. "Presence of Indicator Plant Species as a Predictor of Wetland Vegetation Integrity: A Statistical Approach." *Plant Ecology* 214 (2): 291–302.

Steen, David A., and James P. Gibbs. 2004. "Effects of Roads on the Structure of Freshwater Turtle Populations." *Conservation Biology* 18 (4): 1143–48.

Steinke, Tom. J. 1988. "Restoration of Degraded Salt Marshes in Pine Creek, Fairfield, Connecticut." In *Proceedings: Fourth Wetlands Conference,* edited by Michael W. Lefor and William C. Kennard. Special Reports 33. Storrs, CT: University of Connecticut.

Stevens, T. H., S. Benin, and J. S. Larson. 1995. "Public Attitudes and Economic Values for Wetland Preservation in New England." *Wetlands* 15 (3): 226–31.

Storey, Kenneth B. 1990. "Life in a Frozen State: Adaptive Strategies for Natural Freeze Tolerance in Amphibians and Reptiles." *American Journal of Physiology-Regulatory, Integrative and Comparative Physiology* 258 (3): R559–R568.

Stuckey, Ronald L. 1980. "Distributional History of *Lythrum salicaria* (Purple Loosestrife) in North America." *Bartonia* 47:3–20.

Sutton-Grier, Ariana E., Amber K. Moore, Peter C. Wiley, and Peter E. T. Edwards. 2014. "Incorporating Ecosystem Services into the Implementation of Existing U.S. Natural Resource Management Regulations: Operationalizing Carbon Sequestration and Storage." *Marine Policy* 43:246–53.

Sweet, William, J. Park, J. Marra, C. Zervas, and Stephen Gill. 2014. *Sea Level Rise and Nuisance Flood Frequency Changes around the United States*. NOAA Technical Report NOS CO-OPS 073.

Takahashi, Mizuki K., and Matthew J. Parris. 2008. "Life Cycle Polyphenism as a Factor Affecting Ecological Divergence within *Notophthalmus viridescens*." *Oecologia* 158 (1): 23–34.

Takahashi, Mizuki K., Yukiko Y. Takahashi, and Matthew J. Parris. 2011. "Rapid Change in Life-Cycle Polyphenism across a Subspecies Boundary of the Eastern Newt, *Notophthalmus viridescens*." *Journal of Herpetology* 45 (3): 379–84.

Teal, John M. 1962. "Energy Flow in the Salt Marsh Ecosystem of Georgia." *Ecology* 43 (4): 614–24.

Teal, John, and Mildred Teal. 1969. *Life and Death of the Salt Marsh*. New York: Ballantine Books.

Thompson, Daniel Q., Ronald L. Stuckey, and Edith B. Thompson. 1987. "Spread, Impact, and Control of Purple Loosestrife (*Lythrum salicaria*) in North American Wetlands." US Fish and Wildlife Service, Northern Prairie Wildlife Research Center Online.

Tiner, Ralph W. 2003. *Correlating Enhanced National Wetlands Inventory Data with Wetland Functions for Watershed assessments: A Rationale for Northeastern US Wetlands*. Hadley, MA: US Fish and Wildlife Service, National Wetlands Inventory Program, Region 5.

Treberg, Michael A., and Brian C. Husband. 1999. "Relationship between the Abundance of *Lythrum salicaria* (Purple Loosestrife) and Plant Species Richness along the Bar River, Canada." *Wetlands* 19 (1): 118–25.

Tripepi, R. R., and C. A. Mitchell. 1984. "Stem Hypoxia and Root Respiration of Flooded Maple and Birch Seedlings." *Physiologia Plantarum* 60 (4): 567–71.

Turner, R. Eugene, Ann M. Redmond, and Joy B. Zedler. 2001. "Count It by Acre or Function—Mitigation Adds Up to Net Loss of Wetlands." *National Wetlands Newsletter* 23 (6): 5–6.

Ullman, Roger, Vasco Bilbao-Bastida, and Gabriel Grimsditch. 2013. "Including Blue Carbon in Climate Market Mechanisms." *Ocean & Coastal Management* 83:15–18.

US Department of Agriculture. 2005. *Fire Effects Information System: Rapid Assessment Reference Condition Model; Atlantic White Cedar Forest*. https://www.fs.fed.us/data base/feis/pdfs/PNVGs/Southeast/R9AWCF.pdf.

———. n.d. "Introduced, Invasive, and Noxious Plants." Plants Database. Accessed September 14, 2018. https://plants.usda.gov/java/noxComposite.

US Energy Information Administration. 2017. "How Much Carbon Dioxide Is Produced by Burning Gasoline and Diesel Fuel?" May 19. https://www.eia.gov/tools/faqs/faq.php?id=307&t=10.

US Fish and Wildlife Service. 2005. "Recovery Plan for Vernal Pool Ecosystems of California and Southern Oregon." Region 1, Portland, Oregon. https://www.fws.gov/sacramento/es/Recovery-Planning/Vernal-Pool/.

US Global Change Research Program. 2014. "National Climate Assessment." http://nca2014.globalchange.gov/report/our-changing-climate/.

Vasconcelos, Daniel, and Aram J. K. Calhoun. 2006. "Monitoring Created Seasonal Pools for Functional Success: A Six-Year Case Study of Amphibian Responses, Sears Island, Maine, USA." *Wetlands* 26 (4): 992–1003.

Venne, Louise S., Jo-Szu Tsai, Stephen B. Cox, Loren M. Smith, and Scott T. McMurry. 2012. "Amphibian Community Richness in Cropland and Grassland Playas in the Southern High Plains, USA." *Wetlands* 32 (4): 619–29.

Vileisis, Ann. 1997. *Discovering the Unknown Landscape: A History of America's Wetlands*. Washington, DC: Island Press.

Wacker, Tim. 2007. "Seeking Cause and Cure for Ailing Wetlands." *New York Times*, July 15. https://www.nytimes.com/2007/07/15/nyregion/nyregionspecial2/15Rmarsh.html.

Wallace, J. Bruce, and Joe O'Hop. 1985. "Life on a Fast Pad: Waterlily Leaf Beetle Impact on Water Lilies." *Ecology* 66 (5): 1534–44.

Wallis, Robert C. 1960. *Mosquitoes in Connecticut*. Bulletin 632. New Haven: Connecticut Agricultural Experiment Station.

Wamsley, Ty V., Mary A. Cialone, Jane M. Smith, John H. Atkinson, and Julie D. Rosati. 2010. "The Potential of Wetlands in Reducing Storm Surge." *Ocean Engineering* 37 (1): 59–68.

Warren, R. S. 2014. "Salt Marshes and Sea Levels in Eastern Long Island Sound." *Narragansett Bay Journal* 28. Narragansett Bay Estuary Program.

Waterway, Marcia J., Takuji Hoshino, and Tomomi Masaki. 2009. "Phylogeny, Species Richness, and Ecological Specialization in Cyperaceae Tribe Cariceae." *Botanical Review* 75 (1): 138.

Watson, E. B., A. J. Oczkowski, C. Wigand, A. R. Hanson, E. W. Davey, S. C. Crosby, R. L. Johnson, and H. M. Andrews. 2014. "Nutrient Enrichment and Precipitation Changes Do Not Enhance Resiliency of Salt Marshes to Sea Level Rise in the Northeastern US." *Climatic Change* 125 (3–4): 501–9.

Werner, K. J., and J. B. Zedler. 2002. "How Sedge Meadow Soils, Microtopography, and Vegetation Respond to Sedimentation." *Wetlands* 22 (3): 451–66.

Wessels, Tom. 1997. *Reading the Forested Landscape: A Natural History of New England*. Woodstock, VT: Countryman Press.

Whigham, D., and R. L. Simpson. 1992. "Annual Variation in Biomass and Production of a Tidal Freshwater Wetland and Comparison with Other Wetland Systems." *Virginia Journal of Science* 43 (1A): 5–15.

Whitaker, John O. Jr. 2004. "Sorex cinereus." *Mammalian Species* 743:1–9.

Wigand, Cathleen, Charles T. Roman, Earl Davey, Mark Stolt, Roxanne Johnson, Alana Hanson, Elizabeth B. Watson, S. Bradley Moran, et al. 2014. "Below the Disappearing Marshes of an Urban Estuary: Historic Nitrogen Trends and Soil Structure." *Ecological Applications* 24 (4): 633–49.

Willey, Neil. 2016. *Environmental Plant Physiology*. New York: Garland Science, Taylor & Francis.

Wilson, Edward O. 1984. *Biophilia*. Boston: Harvard University Press.

Winter, Thomas C., and Donald O. Rosenberry. 1998. "Hydrology of Prairie Pothole Wetlands during Drought and Deluge: A 17-Year Study of the Cottonwood Lake Wetland Complex in North Dakota in the Perspective of Longer Term Measured and Proxy Hydrological Records." *Climatic Change* 40 (2): 189–209.

Woodhouse, William W. Jr., Ernest D. Seneca, and Stephen White Broome. 1972. *Marsh Building with Dredge Spoil in North Carolina*. Bulletin no. 445. Raleigh: North Carolina Agricultural Experiment Station.

Zedler, Joy B. 2003. "Wetlands at Your Service: Reducing Impacts of Agriculture at the Watershed Scale." *Frontiers in Ecology and the Environment* 1 (2): 65–72.

———. 2004. "Compensating for Wetland Losses in the United States." *Ibis* 146 (s1): 92–100.

———. 2014. "A *Wet*land Ethic?" Leaflet 36. University of Wisconsin Arboretum. https://arboretum.wisc.edu/content/uploads/2015/04/36_ArbLeaflet.pdf.

Zedler, Paul H. 2003. "Vernal Pools and the Concept of 'Isolated Wetlands.'" *Wetlands* 23 (3): 597–607.

Zwinger, Ann. 1970. *Beyond the Aspen Grove*. New York: Random House.

Index

Page numbers in italics refer to boxes and tables.